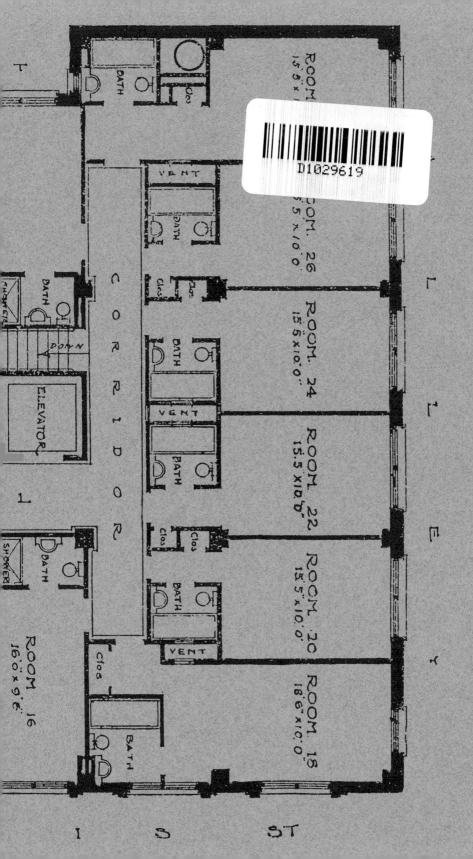
D1029619

THE WINECOFF FIRE

The Untold Story
of America's Deadliest
Hotel Fire

THE WINECOFF FIRE

The Untold Story
of America's Deadliest
Hotel Fire

Sam Heys
Allen B. Goodwin

LONGSTREET PRESS
Atlanta, Georgia

Published by LONGSTREET PRESS, INC.
A subsidiary of Cox Newspapers,
A division of Cox Enterprises, Inc.
2140 Newmarket Parkway
Suite 118
Marietta, GA 30067

Copyright © 1993 by Sam Heys and Allen B. Goodwin

All rights reserved. No part of this book may be reproduced in any
form by any means without the prior written permission of the
Publisher, excepting brief quotations used in connection with reviews,
written specifically for inclusion in a magazine or newspaper.

Printed in the United States of America

1st printing, 1993

Library of Congress Catalog Number 92-84011

ISBN: 1-56352-069-9

This book was printed by R.R. Donnelley & Sons,
    Harrisonburg, Virginia
The text was set in Bembo

Cover design by Tonya Beach
Book design by Laurie Shock

DEDICATION

To Leary Bell Finley
and
Elizabeth Goodwin

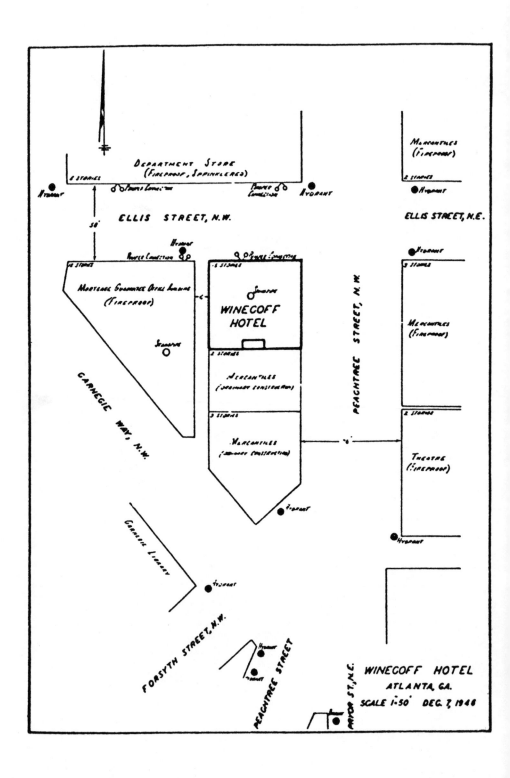

CHAPTER 1

A t three o'clock in the early morning of December 7, 1946, Alice Edmonds was sitting in the third-floor linen closet looking at a comic book. During the ten weeks she had been the night maid at the Winecoff Hotel — at the corner of Ellis and Peachtree, downtown Atlanta — she had spent much of her shift each night in the third-floor closet, awaiting a call to clean a room. Already, Comer Rowan, the overnight desk clerk, had told her to go to Room 616 to change the linens. And Bill Mobley, the bellhop, had sent her for sandwiches for a couple in 1618.

After bringing back the sandwiches sometime around 1:00 a.m., Edmonds completed one of her primary responsibilities: an inspection of the entire hotel. Starting on the top floor, she worked her way down, picking up trash and making sure the linen closet on each floor was locked and the hall clear of obstacles — in case of fire. The Winecoff advertised itself as "fireproof," but its employees knew it wasn't. Only the shell of the brick building would not burn; the contents would, just as they would in any hotel in America in 1946. But unlike many hotels, the Winecoff, built in

1913, had neither fire escapes nor a sprinkler system. Atlanta law did not require them.

At fifteen stories, the Winecoff was Atlanta's tallest hotel, and by the time Edmonds had made her way down from the sixteenth floor to the lobby — there was no thirteenth floor — it was time for her supper break. She walked a block down Peachtree Street to Hamburger Heaven and ordered two hamburgers and cup of coffee to go. When she returned to the hotel, it was two o'clock, time for elevator operator Rozena Neal to take her supper break. So instead of enjoying her own food, Edmonds took over the elevator for Neal's thirty-minute break. Although its 195 rooms were again filled to capacity, the hotel was fairly quiet. More than 280 registered guests were spending Friday night at the Winecoff, but most of them had plans for Saturday and were already in bed: doctor's appointments, a full day of Christmas shopping, a return trip home after a week of work in Atlanta.

The hotel was also full of teenagers who had come to Atlanta for the Youth Assembly at the Capitol. They were from high schools across Georgia, members of the Hi-Y or Tri-Hi-Y, Christian service clubs affiliated with the YMCA. The Y was sponsoring a mock legislature, and the delegates were staying at Atlanta's downtown hotels. Forty had checked into the Winecoff Friday, bunking three and four to a room. Some had brought sleeping bags, and the Winecoff had put roll-away beds into the rooms of others.

The teenagers had spent the evening reveling in the experience of being with their friends in a big hotel away from home. They had been in the lobby and up on the mezzanine, talking, making new friends. A cluster had stood on the mezzanine outside the Plantation Room, the hotel's bar. They had told the lounge's manager, Andy Marnas, that they wanted to sing a few songs with pianist Ward Duvall. Marnas invited them in, and the group sang a couple of show tunes — "Showboat" and "Only Make Believe" — before heading off to their rooms.

They were all in bed by the time Edmonds took over the elevator for Neal. Edmonds took some guests coming in from a late night on the town to their rooms and made several trips between the lobby and the third floor. She knew there was a big card game still going on in Room 330.

Her linen closet was adjacent to the room, and she had heard loud talking and swearing coming through the wall. At one point, Edmonds took three men from 330 down to the lobby. One of the men was short and wore a blue overcoat. Another was tall and heavy. The third had a scar on his face. The trio returned to the Winecoff a short time later with a teenager in a beige overcoat and white pants. Edmonds took them up to the third floor.

When Neal returned to the elevator from her supper break around 2:30, Edmonds went back to the third-floor linen closet to eat her hamburgers and drink her cold coffee. The talking in 330 got louder and louder. Finally closing the door to get some peace and quiet, she picked up the comic book again.

At 2:45, overnight telephone operator Catherine Rowan, the desk clerk's wife, took a call from Warren Foster in 508. Foster, a young father from Columbus, Georgia, was complaining about the ruckus in the room next door. The noise had awakened his baby girl again. Comer Rowan told Mobley, the bellhop, to go quiet the people down. Mobley had already been up to Suite 510-12 once since coming on duty shortly before midnight, to answer a room-service call for ice and ginger ale from two soldiers. This time, Mobley took them more ice for their drinks. When he told the soldiers they'd have to be quiet because other guests were complaining, one of them left.

About a half-hour later, Mobley prepared to make his mid-shift inspection of the hotel. Like Edmonds, he would start at the top of the building and walk every floor, looking for unlocked doors, running water, or smoke. Fred Grennor, the overnight engineer who had started work at the Winecoff just the night before, would accompany him.

Before Mobley and Grennor could take the elevator to the sixteenth floor, Comer Rowan took yet another call for room service from Harry Lindervall, the remaining soldier in Suite 510-12. He wanted more ice and ginger ale. So on their way up, Mobley and Grennor stopped by his suite.

Lindervall took two to three minutes to open the door. When he finally did, Mobley went in to refill the ice bucket while Grennor waited in the sitting room, nervously smoking a cigarette. He was anxious to get on with the inspection. But when Mobley and Lindervall came back into the sitting room, they both had drinks and offered one to Grennor.

He declined.

While Mobley and Grennor were in 510, Rozena Neal took some guests to their room on the tenth floor. On her way back down to the lobby, somewhere around the fourth or fifth floor, she was jerked alert by the smell of smoke. She rushed to tell Rowan as soon as she reached lobby. He told her to go find Mobley and then glanced at the panel above the other elevator. "He's on the fifth floor," he yelled. "Now hurry!"

When the elevator doors opened on the fifth floor, the hallway was already filled with smoke. Neal hollered for Mobley, but the thickening smoke forced her back into the elevator before she heard any answer. She closed the doors and headed for the lobby. As she passed the third floor, she could see flames through the windows of her car. Panicking, she sent the car all the way to the basement, where she started shrieking: "Fire! Fire! Fire!"

Up in 510, Mobley, Grennor, and Lindervall heard a terrifying scream from somewhere down the hall. Mobley stepped quickly to the door. When he opened it, smoke poured in, burning his eyes. He squinted down the hall and saw a mass of flame coming up the stairs. The flames jumped to the walls and sped toward him like a ball off a bat. He lunged back into the room, slamming the door behind him. He immediately called the front desk, but the phone rang and rang without an answer.

After Neal yelled at him from the elevator, Comer Rowan had thought first of his wife up on the fourth floor. She had been feeling sick all evening and had gone to use the bathroom in 410, a room reserved for a late arrival.

"Get back down here quick," he called and told her.

She ran to the door, but when she opened it, flames jumped up in her face and singed her hair. She slammed the door and ran to a window.

Rowan sprinted up to the mezzanine to determine if the hotel was indeed on fire. He looked up the stairs toward the third floor, saw the flames, and ran back down to the front desk to phone the fire department. Located only 800 feet away, Station 8's trucks would be on the scene in a matter of seconds. Rowan then began to call the rooms of other Winecoff employees. His message was terse: The hotel is on fire. Keep your door closed to prevent a draft from spreading it.

Alice Edmonds was trapped in the linen closet on the third floor. She had smelled smoke and opened the closet door, stepped into the front hall, and looked across the T-shaped stairway. She saw a huge ball of smoke in the back hall. It had a funny odor to it, she thought, like gasoline or burning tires. She impulsively closed herself back in the closet and screamed, over and over, until there was no air left in her lungs. Then she realized she had to escape. She reopened the door and tried to go down the stairs, but she couldn't even get to the landing. The smoke drove her back to the third floor. Screaming again, she pounded on the door of 304.

"Let me in, let me in!" she shouted.

Two young soldiers opened the door. She stumbled into the room and fell on the floor. Her face was burned, and so were her hands and arms and legs. She was sobbing uncontrollably.

"Lord, have mercy on my soul," she cried.

CHAPTER 2

— ≡✦≡ —

F riday had been a long, frustrating day for Jimmy Cahill. He spent most of it at Georgia Tech, a mile north of the Winecoff. He waited four hours to see the dean, along with a room full of other returning GIs, all seeking the same goal — to return to school.

That had always been Cahill's intention. For the previous five years — while waiting to make bombing runs in the Pacific, standing in line, or trying to occupy all the other dead time of the war — he had planned his life to come. He would go back to Georgia Tech, finish his education, and earn enough money to support a family. All he had to do was make it back alive.

Having survived fifty-three missions as the navigator of a B-25, Cahill — along with his wife, Doris — had driven across the country to his home in Albany, Georgia, in time for Thanksgiving. Although not scheduled for discharge until January, Cahill had accumulated so much leave time that he was actually on "terminal leave." He was already starting his new life. He drove to Atlanta Thursday, over the same route he had taken a decade earlier, right out of high school, when he had come to Georgia Tech to play football. This time he was coming to buy civilian clothes — returning veterans were snatching them up as quickly as they could be manufactured — and to reapply to Tech.

He hardly received a hero's welcome. When the dean finally agreed to see him, he gave Cahill little hope. He talked mostly about Cahill's poor academic record before the war, when he had flunked out of Tech as a freshman and had to go to evening school to be readmitted.

"I know what I did here before I was drafted," Cahill told him. "But I was a seventeen-year-old kid. I'm a grown man now. I've been in the service for six years. I've had positions of responsibility. I'm a major in the Air Corps."

The dean was unimpressed. The war was ancient history to him.

"A leopard never changes his spots," he replied.

Cahill was seething. "What kind of country did I risk my life for?" he wondered. "Doesn't this guy know what we did over there?"

But Cahill was also desperate. Although he had always planned to return to school, the importance of a college degree had increased significantly a few days earlier. Doris had told him she was pregnant. Now he *had* to get a degree.

The dean gave Cahill an option: go to evening school for an indefinite period and hope eventually to be readmitted to Tech.

"My best advice to you," the dean said finally, "is to get a job digging ditches."

The words burned in Cahill's mind the rest of the day. They were still ringing in his ears when he went to bed in Room 608. A few fitful hours later, he woke up to screaming. He figured it was just the teenagers in the hotel, whooping it up. He picked up the phone to complain to the front desk, but the wires had already been burned, and the line was wide open. He could hear nothing but cries for help from other guests. Some were pleading desperately, others screeching in panic. Then he realized he smelled smoke. As he shook Doris awake, he immediately thought of his mother, Bessie Tarver, who had come with them to shop and whose room, 620, was less than twenty-five feet away.

Cahill jumped up and quickly put on his pants and shoes. But after opening the door and seeing a wall of smoke and flame, he knew he wouldn't be able to get down the hall to his mother. He quickly decided he would first have to escape the hotel himself, then get his mother out. Making snap decisions in the face of the utmost danger was not a new experience for Cahill. Navigating a B-25 through enemy fire in the Philippines, New Guinea, and New Britain had accustomed him to thinking under life-and-death pressure, as well as to the sight of smoke and fire.

After Cahill and his wife finished dressing, they threw their luggage out the window and then slammed it shut, trying to stop the draft that was sucking smoke through the cracks around their door and transom. But the smoke was already so thick that they had to reopen the windows and stick their heads out into the night air.

Peachtree Street spread out below them. A few pedestrians had gathered on the sidewalks, but no fire trucks were in sight. They saw other guests in the windows, coughing and crying for help. Three floors below them and two rooms to the right, they saw a woman sobbing hysterically, as if in shock.

It was Alice Edmonds, the night maid, who had first smelled smoke while sitting in the third-floor linen closet. Her screams had awakened the Casey brothers from Guthrie, Oklahoma, who were sleeping across the hall in 304. They first thought the cries were those of a woman being beaten and screaming with every breath. The linen closet was only eight feet away, but Bill Casey saw no one when he cracked the door to look into the hall — only white smoke. It smelled like coal smoke to him, coal smoke with something else mixed in. He wasn't sure what. He closed the door, but the screams continued. Then there was heavy pounding on the door.

"Let me in, let me in!" shouted the voice on the other side.

When the Caseys jerked the door open, Edmonds stumbled into the room and fell to the floor. The brothers picked her up and took her to the windows.

Bill Casey was a twenty-seven-year-old Air Corps sergeant stationed at Shaw Field outside Columbia, South Carolina. He had caught a bus to Atlanta Thursday night, arrived at dawn Friday, and headed to the Atlanta Ordnance Depot at Conley, just south of town, where his brother Pat, a twenty-one-year-old Army T/4, was completing a course. Pat was scheduled to ship back to Camp Hood in Texas over the weekend, and Bill knew this would be their last chance to be together for a while.

The Caseys caught a bus downtown late Friday afternoon and went to Blick's bowling alley on Ivy Street, a block off Peachtree Street. After bowling a few games and drinking a beer, they checked into the Winecoff and then headed back out to find a movie.

There were four movies within a block of the hotel; that was one of the attractions of the Winecoff. A block north on Peachtree, Irving Berlin's *Blue Skies* was in its second week at the Roxy, and Hopalong Cassidy was playing in a double-feature Western next door at the Capitol. Diagonally across Peachtree from the Winecoff, the Loew's Grand had Katharine Hepburn in *Undercurrent*, and the Paramount was showing Atlantan Joel Chandler Harris's *Song of the South*. Just three blocks away, around the corner and down Forsyth Street, was the Rialto and *Three Little Girls in Blue*.

The Caseys chose the Hopalong Cassidy double feature. They returned to the Winecoff about eleven, had a couple of drinks, and were asleep before midnight.

When the first fire truck arrived, the Caseys hollered down that Edmonds was burned and to send up a ladder quickly. A fireman yelled back that someone would be coming up the stairs to get them. Within a few minutes, they heard a crashing thud behind them. They turned and saw that a fireman had knocked the paneling out of their door.

"You can go down the stairs," he said.

As the Caseys helped Edmonds down the steps, they paused to let another fireman go down ahead of them. He was carrying the body of a teenage girl. Her face and nightgown were bloody. Another fireman carried the body of a girl who appeared unconscious. The firemen had scooped up both teenagers from a small, submerged concrete court outside the windows of Room 330, where the big poker game had been raging only a half-hour earlier. The poker players had climbed out the window onto the concrete court area and then up to the roof of the adjacent Chandler's Shoe Store. As they were escaping, bodies were falling from the rooms above. A fireman had noticed the girls outside the window when he checked to see if anyone was in 330.

—•—≡◆≡—•—

Below him and to his left, Cahill could see bellhop Bill Mobley and night engineer Fred Grennor, who had been trapped in Harry Lindervall's suite, 510-512. Lindervall was frantic. He had thrown a white packet to the sidewalk and was yelling to the people on the street not to bother it, that it contained official government papers.

When flames burned down the hall door and then started through the door between the sitting room and bedroom, Mobley, Grennor, and Lindervall climbed out the window. Just outside was a steel cable supporting the tall, lighted "Winecoff" sign that was braced vertically to the hotel's northeast corner, so as to be visible on both Peachtree and Ellis streets. Grabbing the cable, the men lowered themselves one floor to an aerial ladder. Lindervall was so drunk that Grennor and Mobley had to help him down.

Directly below Cahill in Room 508, Warren Foster — the man who had complained to the front desk about the disturbance in Suite 510-512 — watched as the three men got on the ladder. Foster had been back asleep for less than an hour when smoke filled his room. It was so thick he couldn't even see the light he had left on in the bathroom. His wife, Margaret, was so overcome by the smoke that he had to shake her twice to wake her. When he cracked the door, the noise in the hall sounded like a locomotive. Words he had heard fewer than ninety minutes earlier — words from a man storming out of 510-512 — flashed through his mind: "I'll get even with you before the night's over."

Foster had checked into the Winecoff Thursday for a short but needed vacation. During the war he had helped convoy supplies to England for D-Day, and it had been a very trying year since he had come home from the Navy. He and his wife had bought a grocery store in Columbus, and business had been good. But they had been held up at gunpoint once, and Margaret, only twenty-one, often had to stay at the store by herself when Warren went to the market to buy produce. So they had put the store up for sale, and when it was finally purchased, they had decided to take a vacation.

Friday evening the Fosters went to College Park to visit friends and returned around 12:30. They heard the men next door return to 510-512 right after that. It sounded as if there were three of them.

First it was the men's voices that woke up little Connie, who had turned one year old on Thursday. Every time Warren and Margaret would get her back to sleep, the loud talking would awaken her again. An adjoining door between 508 and 510 — the Fosters' room could be added to the suite — enabled Warren and Margaret to hear the men's conversation clearly. One of them accused another one of flirting with someone downstairs in the bar. Finally the arguing escalated into a scuffle. The Fosters heard glass breaking, and Warren could tell punches were being thrown. Furious that his baby couldn't sleep, he called the front desk.

Within minutes, he heard the bellboy — Mobley — knocking on the door of 510-512 for the first time, trying to quiet the men. Afterward he heard the threat from the man leaving the room.

Foster was still thinking about the threat as he stood at the window with Margaret, her arms outstretched, holding Connie in the night air.

"Get that baby!" hollered the people who had already gathered on Peachtree Street. The firemen quickly moved the aerial ladder to the Fosters' window.

⚜

As the Fosters climbed down carefully, Cahill leaned out of the window directly above them and started barking orders to the firemen on the street. He was anxious for them to rescue the hysterical young couple next door to him in 606. Only minutes earlier, the guests in the room two floors above Cahill had yelled for the couple to wrap wet towels around their heads.

"They can't get to the bathroom," Cahill shouted back. "Their room's on fire. I can see the flames."

The guests on the eighth floor dropped wet towels and Cahill caught them, but he wasn't able to get the couple's attention, so he leaned as far as he could out his own window and popped the young man in the face with the towel. Dazed by smoke and fear, the man still didn't respond. Cahill popped him again, then again — three times Cahill smacked him before the man took the wet towels and began to breathe through them, getting oxygen from the water.

After the couple had been rescued, Cahill shouted to the firemen, "Now move that ladder over here."

The ladder was a foot short of reaching the Cahills' window, but that was close enough for them. Cahill helped Doris climb out and then followed right behind her, stepping on her hands several times. He was in a hurry. Once on the ground, Cahill got Doris out of the way of danger and then ran around the corner of the Winecoff and down Ellis Street. The forty-four-degree air quickly cooled his body, which was still perspiring from the furnace-like heat in 608. All the anger he had felt before going to bed was gone. His mind was totally focused on how he would get his mother out of her room. He didn't even hear the screams of the two women perched on a ledge forty feet above him.

The two women, Lib Stamps and Catherine Weliver, had checked into Room 418 Friday, after a long train ride from Crawfordsville, Indiana. Catherine's husband, Otis Weliver, managed the Coca-Cola bottling plant there and was attending lab school at the company's headquarters in Atlanta. Earlier in the week, he had asked Catherine to join him, and she invited her cousin Lib to come with her. They were at the Winecoff because Weliver had been unable to get them a room at the Biltmore, where he was staying.

The two women awoke to yelling in the hall. Catherine Weliver thought it was a fight. But when she opened the door, she saw people sprinting through flames toward the stairs. Suddenly flames jumped up in her face and singed her hair. She slammed the door.

The room quickly filled with smoke and heat, forcing Weliver to the window. Stamps was already hanging out of the bathroom window. She had been wetting towels and was unable to reenter the bedroom, repelled by the smoke that had rushed in through the briefly opened door. From the two windows the women saw each other — and the ledge. It was actually an eight-inch-wide cornice that topped a decorative tile facing on the Peachtree Street and Ellis Street sides of the Winecoff.

The firemen from Station 8 were running up an aerial ladder to the guests on that side of the hotel. Other firemen were holding nets for the guests who were ready to jump.

"I'm going to jump, Lib," hollered Weliver, a thirty-four-year-old mother of two children, after the two women climbed onto the narrow ledge.

Stamps could see that her cousin was becoming hysterical. She, too, was terrified but was fighting to remain calm.

"Oh, don't do that," she replied. "I wouldn't want to have to stay up here by myself."

The women waited and waited — it seemed an eternity — for the aerial ladder to get to them. When it finally did, they quickly descended to Ellis Street. Then the men from Station 8 cranked their big ladder toward the fifth floor and Emogene and Gene Bates.

Just like Catherine Weliver in the room directly below her, Emogene's hair had been singed when she opened her door. The hall looked like an inferno. She ran to the bed and shook her eleven-year-old son, Gene.

"You gotta wake up. The hotel's on fire," she told him.

Room 518 quickly filled with smoke, forcing the Bateses to wrap towels around their heads. Trying to get away from the smoke, Gene fled to the bathroom and lay down in the tub. It was easier to breathe there, but his mother hadn't seen him slip away and didn't know where he was. She looked frantically for him. When she found him, she took him to the window. There, while trying to breathe through the black smoke being sucked out of the building, thirty-eight-year-old Emogene Bates wondered if this was what dying felt like. She thought about her husband, Glen, the assistant police chief in Murphy, North Carolina, who had stayed at home to work while she and Gene came to Atlanta to Christmas shop. When the fire burned through the door and flames entered the room, she asked her son, "Do you think we should jump?"

His answer was quick: "No."

But sensing his mother's increasing panic, Gene began yelling as well, hoping the firemen would realize he was a child and move the ladder to their window next. They did.

As Gene made his way down — slowly and carefully, step by step — he saw a woman fall from a window above him. It seemed she was desperately reaching for the ladder.

Once the Bateses were on the ground, the firemen quickly swung the ladder over to the adjacent room, toward the terrified, piercing screams of Lillian Ahern, who had started yelling as soon as she and her husband had awakened and seen smoke pouring through the transom.

"It's going to be all right," Hugh Ahern said repeatedly. But Lillian wouldn't stop screaming.

The Winecoff had been the Aherns' home for more than a week. Hugh had been transferred to Atlanta from Washington by the U.S. Department of Commerce, and they were waiting to move into the house they had bought. They had spent their first night in the hotel on the eleventh floor, in a room on the back side that was barred with exterior blinds. "We've got to get another room," Lillian told her husband before going to bed. "If this place caught on fire, we'd never get out of here."

The possibility of fire was never far from the minds of guests staying in postwar America's 1.5 million hotel rooms, for the United States had been haunted by fire during the first half of the twentieth century. The nation was changing — its population migrating from the country to the city — but the law was not keeping pace. The horror of high-rise fire had struck as early as 1911 when flames erupted on the upper three floors of the ten-story, "fireproof" Asch Building, killing 145 employees of the Triangle Shirtwaist Company. More recently, inadequate fire prevention had been blamed in 1942 when 491 people died at the Coconut Grove nightclub in Boston. Two years before that, 198 people had been killed in a dance-hall fire in Natchez, Mississippi. And in 1944, 168 people, mostly women and children, died when fire broke out in a Ringling Brothers Circus tent in Hartford, Connecticut.

After their first night at the Winecoff, the desk clerk had asked the Aherns if they would like to move to 516. He warned them it was a small room, with a shower instead of a bathtub and a wardrobe instead of a closet. Winecoff employees referred to the rooms facing Ellis Street — numbered "14" or "16" — as "commercial rooms" because they were usually rented to salesmen traveling alone and trying to save a buck. The Aherns took the room.

Lillian Ahern knocked out the screens of 516 with her umbrella. When Station 8's aerial ladder was jammed against the bricks between the room's two windows, she realized she was about to be separated from her husband. She feared she would be saved only to lose him.

"I'm not going unless you promise me you'll come right back and get my husband," she told the fireman who came up the ladder.

He promised.

As Alvin Millman watched the Aherns go carefully down the ladder, he believed he would be next. The thirty-eight-year-old Chicago businessman was staying next door in 514. He had wisely not opened his door, having read repeatedly about the La Salle Hotel fire in his hometown in June.

With sixty-one deaths, the La Salle had been the second worst hotel fire in U.S. history, its death toll having been surpassed only by the seventy-one victims at the Newhall House in Milwaukee in 1873. The blaze was apparently started by a short circuit in a false ceiling above the La Salle's bar. Within minutes, seventeen floors were in flames. Four days later, nineteen people were killed in the Canfield Hotel fire in Dubuque, Iowa. Later that month ten people died in a fire at the Baker Hotel in Dallas.

Once the Aherns were on Ellis Street, Millman saw to his horror that the big ladder was moving away from him into the smoky darkness, headed up to the next floor. With plaster falling all around him and flames licking under the door, he quickly threw his suitcase out the window. That got the firemen's attention. But instead of moving the ladder to Millman's window, they stretched a net out below him. It was fifty feet away. Straight down. He jumped.

He landed flat on his back. His momentum was so great that his back still hit the pavement as the net caught him. But he got up and walked away.

With smoke enshrouding the building, the big aerial ladder also missed C. E. Doyle. Sales manager for Wheatley Mayonnaise Company in Jacksonville, Florida, he was huddled in his overcoat and hat on the window ledge of 616. When the ladder stopped at the room above his, it was only five feet away, and Doyle knew it might not get any closer. He didn't think the firemen could see him, and if they could, maybe they were expecting him to jump into a net. And so many people were crying for help above him. He looked down and saw the flames creeping nearer. He lunged for the ladder.

Doyle's hands caught a rung from the underside. For a second his feet and legs dangled freely; then he pulled them to a lower rung. He climbed down to the street on the underside of the ladder.

*   *   *

By the time Cahill reached the Winecoff's back corner on Ellis Street, the narrow alley behind the hotel was littered with suitcases. Guests on the lower floors, seeing the ladders approach, were throwing their luggage out the windows to save their possessions. Those on higher floors jettisoned their suitcases in a desperate attempt to attract the attention of firemen below.

Cahill stared up at the tall building, searching for his mother. He saw people hanging out of windows, then a body falling through the thick smoke that was funneling upward between the Winecoff and the Mortgage Guarantee Building. But he could not see his mother. Realizing there would be no aerial-ladder rescues for her or for the other guests on the back side, Cahill knew the ten-foot-wide alley would eventually become a death trap for all of them. They would have to pray they could hold on until the fire was extinguished. Or else jump. Suddenly, Cahill started running again, further down Ellis Street. He rounded the corner at Carnegie Way and headed for the entrance of the Mortgage Guarantee Building, a twelve-story office building across the street from the city library. To Cahill, getting into the Mortgage Guarantee Building was the obvious solution, the only solution. It was directly across the alley from the Winecoff and nearly as tall, meaning that forty percent of the hotel's guests were only ten feet from safety — if they could just somehow cross the alley. The fire department had even listed the Mortgage Guarantee Building as an escape route from the Winecoff in its compilation of evacuation plans for every major building in the city. But no one in the fire department had remembered those plans yet.

Cahill was prepared to break the lock or shatter the glass on the front door, but it was already open. He sprinted up the stairs. After about three flights, he began to stop on each landing, throwing up a window, looking and listening, searching for his mother, straining to hear her cries. The smoke above the alley was so dense that Cahill couldn't even see the windows of the Winecoff. Only an occasional gust of wind allowed him a brief glimpse. Finally he heard his mother's voice. She was to his left and just above him. He yelled at her. She shouted back that she could hear him.

"Keep talking so I'll know where you are," he hollered.

Cahill raced up another flight and hurried down the hall. His goal was to get slightly above his mother, as the floors in the Winecoff did not line up precisely with the floors of the Mortgage Guarantee Building. When he came to the office that he thought was across from his mother's window, he found the door locked. As he backed away from the door, preparing to charge it and break it down, a man emerged from the elevator down the hall. He was Charles Burton, chief engineer of the Mortgage Guarantee Building.

Burton had rushed downtown when he heard of the fire, knowing he might be able to help just as he had nearly five years earlier. On February 19, 1942, a fire had broken out on the eleventh floor of the Winecoff when a guest supposedly threw a lighted cigarette or match into a wastepaper basket. No one had been killed or injured, largely because Burton had placed boards across the alley and rescued several guests.

"What are you doing?" Burton asked Cahill.

"I'm getting ready to kick this door in unless you have a key."

"Why?" asked Burton, hurrying down the hall.

"Because my mother's in the hotel, right opposite this office, and somehow I'm going to get her out."

Burton got out his keys and opened the door. Cahill went immediately to the window and looked. There, through the smoke, slightly below him, was his mother, Elizabeth Tarver, crouched on the window ledge of Room 620. She seemed almost close enough to reach out and touch.

"I want a plank to go across this alley," Cahill said. "I'm going to get her."

"I've got a painter's footlog down in the basement," said Burton.

"Hang on!" Cahill yelled to his mother. "I'll be back in a minute or two. I'm going to get you."

As Cahill and Burton ran out of the office and down the hall to the freight elevator, Elizabeth wondered if she could hold on another couple of minutes. Before Cahill had called to her, she had been hoping a falling body would knock her off her perch. She wanted to jump but couldn't make herself. She feared she'd burn to death if she didn't.

Sitting sideways on the ledge, she was leaning hard on the window she had shut behind her so draft-pulled smoke wouldn't strangle her. She had two towels, one was wrapped around her face and the other between her arm and the window so she wouldn't be burned from the heat of the glass. She was repeatedly glancing back to see if the fire had entered her room.

That possibility terrified the guests all across the lower floors of the Winecoff's back side. Their rooms' contents were hardly fireproof: wooden furniture, upholstered chairs, draperies, wooden Venetian blinds. Following the Winecoff's last reported fire — a blazing mattress in Room 1420 back in February — workers had flameproofed all mattresses, drapes, pillows, and even the carpet. But the treatment was not deterring the fire's rampant spread through the Winecoff's lower-floor halls, where fuel for fire was plentiful. The corridor walls were covered with painted buckram or burlap fabric from the wood baseboard to a chair rail four feet above the floor. From the chair rail to the ceiling, the walls were covered with as many as five layers of wallpaper. Wooden picture molding was stripped along the ceiling.

Firemen inside had already concluded that the fire had probably started on the third floor, and John McFalls and Nolan "Hornet"

Russell, a pair of electric utility linemen from North Carolina sleeping in Room 326, had been two of the first guests to discover it.

McFalls had been hard asleep when he heard Russell calling his name. As he struggled from bed to follow Russell to the window, his throat burned so badly he wondered if he was about to strangle. But outside the window the air was fresh, and looking out, the two men could see no sign of fire, not even any smoke. The alley was quiet, almost eerily so.

"Hornet," said McFalls, "let's go outside and see what's the matter."

Russell and McFalls felt their way through the small, dark room. Russell gathered their clothes in his arms as he went by the tree hook where they had hung them ninety minutes earlier.

"Get our shoes," he told McFalls.

"Forget the shoes," said McFalls, who was having increasing trouble breathing.

Russell unlocked the door and turned the knob. The door sprung open as if someone were pushing on it. The hall looked like the inside of a furnace. The flames were roaring so loudly that they smothered all other noise. Smoke belted the men in the face as it raced for their open window. The heat felt as if it would blister their bare skin.

"Shut it," said McFalls.

"I'm trying. You've gotta help me," said Russell, whose stomach had suddenly turned queasy. The flames made his mind flash back to all the fires the Japanese had set in Manila during World War II. He was still emotionally scarred from the fighting there.

With McFalls also pushing, Russell shut the door. But now smoke permeated the room and the men could neither see nor breathe. They grabbed on to each other as they started the groping, fifteen-foot journey back to the window.

Roommates back in Franklin, North Carolina, just across the Georgia line, McFalls and Russell had been planning this trip since July. The War Assets Administration had offered 2,700 Jeeps for sale at the ordnance depot in Conley. Created in 1941 in preparation for war, the depot was now unloading the war's surplus, and veterans were flocking to Conley from across the South. McFalls came to buy a Jeep for his son. Russell, the veteran, was going to be the purchaser.

The two men had caught a 6:30 bus after getting off work Friday evening. They had arrived in Atlanta at midnight and walked the three blocks from the bus station to the Winecoff. After checking into the hotel's last available room, they ate supper at an all-night diner, window-shopped briefly, then came back to the room. They went to bed about two o'clock.

By the time McFalls and Russell had made it to the windows the second time, the alley had come alive with noise: sirens screeching, windows opening, people screaming.

McFalls was looking to his right at a fire truck pulling up on Ellis Street when he saw something go by out of the corner of his eye. It was a man in his underwear, falling from above. McFalls saw the body strike the pavement of the alley.

The fire truck that McFalls saw was from Station 8, located two blocks away at the corner of Spring Street and Carnegie Way. Two minutes earlier, its men had been asleep. Once awake, they had thought they would be back in their bunks in twenty minutes, that they would only have to extinguish a mattress or kitchen fire. But when they arrived, they saw that the fire had already engulfed the third, fourth, and fifth floors and flames were leaping out the windows. When they turned the truck's engine off, they could hear the screaming.

When the firemen ran into the alley, McFalls and Russell hollered for a ladder. Through the smoke, a fireman stretched one toward them, and Russell reached out the window to grab it. As McFalls followed Russell down — their pants thrown across their shoulders — another body sailed past them, crashing into the alley. It sounded like a dropped watermelon.

"We gotta get out of here, or we're going to get hit," yelled Russell as the two ran barefooted from the alley. Not until they were out on Ellis Street did they stop to put on their pants and shirts.

Back in 324, Paul Lankford was praying. He was certain that a ladder would not come to him and he would soon die. He was asking God to allow him to retain his mental faculties until the end. A twenty-six-year-old measurement supervisor for Southern Natural Gas, Lankford was on assignment in Atlanta and had been living in the Winecoff for a month. He planned to go home to Birmingham on Saturday to see his wife and nine-month-old baby girl.

Lankford's boss, Roy Peace, had come over Friday from Birmingham, and the two of them had gone out to dinner with Peace's brother, Frank. They had returned to the Winecoff just before midnight. On their way up to Peace's room on the fourth floor, three men had boarded the elevator with them. They had obviously been drinking and asked Frank Peace to have a drink with them. He declined. The three got off on the third floor, turned right, and headed around the corner to Room 330, where the poker game was in progress.

Lankford was in bed by 12:15. Around 3:30, he heard someone trying to open his door. He stumbled out of bed, thinking it was the hotel house detective, who had awakened him the night before to tell him his door was unlocked. Angry that he had been disturbed again, Lankford was ready to take a swing at the detective. As he reached the door, he heard running in the hall. He opened it and flames shot into the room. He slammed the door and ran to the window. As he knocked the screen to the alley, he wondered why he hadn't smelled any smoke before he opened the door.

"Fire!" he yelled. "Fire!" Again and again his screams reverberated against the silence of the night. He saw no one.

Within only a couple of minutes, shrieking filled the alley, and the door began to burn through. Lankford pulled the window down across his back, denying the fire any additional oxygen.

Lankford heard someone calling to him from above. He turned and looked up. It was Jesse Carroll in 424.

"I gotta get out of here," yelled Carroll. "I'll fry in this room."

"Well, my room's on fire," hollered Lankford.

"I don't care, it can't be as bad as mine. I'm coming down."

Carroll was twenty-five, a wiry 140 pounds, and just out of the Army. An accountant for Parks Truck and Equipment Company in Knoxville, he had checked into the Winecoff Monday. He was looking forward to a weekend with his wife and two little girls and planned to start the six-hour trip home early Saturday morning.

Although hard of hearing, Carroll had been awakened by Lankford's screams. As he dropped his legs out the window and dug his bare toes into the cold mortar between the bricks, he knew he was counting on Lankford, a man he didn't even know, to save him again.

Lankford stepped up onto the windowsill and pulled the window down from the top. Still squeezing the bricks outside his own window, Carroll dangled his legs down. There was four feet of brick from the top of Lankford's window to the bottom of Carroll's. Lankford grabbed Carroll's ankles. Slowly, Carroll moved his hands down the bricks. Lankford was struggling to maintain consciousness, his head engulfed by the dark, gaseous smoke being sucked out the open window. Carroll bent slightly at the knees and gave Lankford first one arm and then the other. They locked arms, and with a tremendous heave Lankford jerked Carroll through the open top half of the window. Both men fell to the floor.

Gasping for air, Lankford got up and stumbled back to the window. He saw that the alley was in chaos. The firemen were having to dodge suitcases raining from the windows, and the smoke was so thick that they couldn't see the people above them. They ran their ladders up randomly and, like fishermen, waited to see if anyone climbed on. If no one did, they moved the ladder to the next window.

After Lankford caught his breath, he turned back into the room to look for Carroll, who he assumed was still on the floor. But he couldn't find him. Suddenly he realized firemen were in the hall extinguishing the fire. He glanced into the corridor, saw the stairwell four feet from his door, and bolted through the firemen's spray and down the stairs.

As Lankford descended to the mezzanine, he saw that neither the bar nor the dining room had burned. He quickly realized the fire had started on the third floor and he had been one of the first people to discover it. He thought about the running he had heard in the hall before he opened his door.

In the lobby Lankford noticed Comer Rowan, the desk clerk, busily trying to use the telephone, but the calls weren't going through. The switchboard was blinking like a Christmas tree.

Among the employees Rowan was desperately trying to contact was thirty-one-year-old Helen Walker, the hotel's auditor. Walker was four days away from graduating from Atlanta Law School, moving out of the Winecoff, and starting work for a judge in Metter, the southeast Georgia town where her husband was trying to establish a dental practice. She was vice president of the Atlanta Women's Chamber of Commerce, a successful career woman, but she was willing to leave Atlanta to make one last attempt to save her marriage.

A year earlier, J. O. Walker had asked her for a divorce so he could marry an English woman he had met during the war. They remained married, however, and she supported him while he tried to start a practice in Savannah, just as she had during his years of dental school.

As Rowan struggled with the useless telephone, the smoke from the fire was driving Walker out onto the window ledge of Room 528, above the back corner of the alley. She thought she was in a position to be rescued, but flames suddenly shot up from the window below and engulfed her. She fell to the ground.

--- ⚔ ---

On the window ledge outside 620, Elizabeth Tarver was fighting to keep from passing out from the smoke and falling to her death when she again heard her son, Jimmy Cahill, on the other side of the alley. The footlog that he and Burton had found was about twelve feet long and constructed of a series of one-inch-by-two-inch boards nailed together. It was sturdy enough for painters to stand on when using scaffolding to paint windows. Cahill and Burton slid the footlog slowly across the alley. They had to hold it still until the smoke stopped swirling, then inch it even more slowly toward the Winecoff. The closer it got to the window, the more careful they had to be not to knock Elizabeth off.

Once the board was on the window ledge, Burton offered to run get a rope to tie around Cahill.

"We don't have time for that," Cahill replied. "You just keep pressure on the board so it won't move."

The board, perched sixty feet above the alley, sloped downward at a thirty-degree angle. Cahill climbed on it carefully, straddling it like a horse.

Elizabeth had spent the night at the Winecoff often, whenever she brought children to Scottish Rite Hospital. She was a public health nurse in Albany — more or less *the* public health department of Albany — and her compassion was legendary. When she had seen the plight of pregnant black women in rural southwest Georgia in the 1920s, she had responded with both caring and innovation. Because there were no black doctors and because few white doctors would treat blacks, she had trained a group of black women to be midwives. The program became a model, copied around the state.

"I knew you'd come get me," she said to her son when he had made his slow, methodical way across the alley. "But you shouldn't have. You could have been knocked off this board and been killed. You shouldn't have done it."

"Hush," said Cahill, who could see flames at the door of his mother's room. "Get on this damn board, and let's get out of here."

Cahill turned around so that he was facing the Mortgage Guarantee Building. His mother straddled the board behind him, her arms reaching under his armpits to clutch his broad shoulders. At five-feet-seven and 175 pounds, Cahill was a strong, sturdy man. He had been a pulling guard in Georgia Tech's single-wing, and coming off his last Air Corps assignment as a physical training instructor, he was probably in the best shape of his life.

With smoke encircling them, mother and son started back across the alley. Reaching out and grasping the footlog on its sides, he pulled the weight of both of them upward, toward the Mortgage Guarantee Building, where Burton continued to keep the board jammed against the window of Room 620. It was a slow trip, made in a concentration-shattering din. The wailing sirens, the cries for help, and the screams of terror created a cacophony of pandemonium. But Cahill could still hear his mother saying softly behind him, "Now, son, be careful and don't fall." Screens, pocketbooks, and bodies were falling all around them. Somehow, they didn't get hit. Somehow, they made it.

By this time, the firemen from Station 8 had seen what Cahill had done and were following his lead, extending their ladders across to the

Winecoff.

In 624, twenty-seven-year-old Irene Justice had also been waiting at the window. She was a secretary for a tax accountant in the nearby Candler Building. From East Point, she had taken a room at the Winecoff for convenience while attending law school downtown in the evenings. To the firemen she looked as if she were asleep on the windowsill, her head resting on her arm, her eyes closed. But when they touched her, they realized she was dead.

At the end of the sixth floor's back hall, in 628, Mildred Johnson was also hoping for a ladder. She initially had been determined to survive, but as the inevitability of her situation set in, an acceptance of death, not panic, swept over her. Dying would be all right, she thought.

When Johnson could breathe at the window no longer, she tied her bedsheets together and secured them to the mullion between the windows. As she started down toward a ladder, flames leaped from the windows of 528 and burned her sheet rope in two. Her fall was broken — and her life saved — by the other bodies piling up at the end of the alley.

Irene Justice, a secretary by day and law student by night, died waiting for a ladder that never came to Room 624. The charred shutters were intended to provide both privacy and ventilation to the non-air conditioned hotel.

CHAPTER 3

B ob and Billie Cox had planned to leave Little Bob home. He had just turned three in November, and the three-hour trip from Murphy, North Carolina, to Atlanta was an uncomfortable ride over winding, mountainous roads, so the Coxes asked Delilah Chambers to come keep Little Bob at their house that weekend. Chambers worked at the mill that made bedspreads outside of town and picked up extra money baby-sitting.

But at the last minute the Coxes decided to take Little Bob with them — it would be his chance to see Santa Claus. They asked Chambers to come with them, so she could babysit while they went out Friday night with Bob and Pauline Bault. The two couples had once shared a duplex in Murphy, a dozen miles across the Georgia line.

The Coxes had met while working on one of the South's biggest construction projects: the building of Fontana Dam on the Little Tennessee River in western North Carolina. Bob, a physician, and Billie, a secretary, both worked for the Tennessee Valley Authority, created by Franklin Roosevelt to prevent flooding of farmland, generate electricity for the development of southern industry, and create jobs. With 6,000 people on the Fontana construction site, TVA had an obvious need for physicians. In 1942 the Coxes had moved to Murphy because of an acute shortage of doctors there. Cox, thirty-three, was co-owner of the local Petrie Hospital and active in the community, serving as chairman of the polio fund drive scheduled to start in January. Billie taught Sunday school and had organized a Girl Scout troop.

After arriving at lunchtime Friday, the Coxes and Baults spent the afternoon shopping, making sure Little Bob saw the mechanical Santa Claus that went "Ho, ho, ho" in the window of Davison-Paxon's. With Davison's directly across Ellis Street and Rich's six blocks down Forsyth Street, the Winecoff attracted shoppers from across Georgia and the Southeast year-round, but never more than in December. That evening, Chambers kept Little Bob while the couples went to the Henry Grady's Paradise Room. They planned to return to Davison's Saturday to let Little Bob ride the four banks of escalators scheduled to begin operating that morning.

The couples said good night when the elevator stopped at the Winecoff's tenth floor. The Coxes had taken a corner suite, 1002-04, so Chambers and Little Bob could have their own room. They had wanted a room next to the Baults, but the desk clerk had told them they would have to wait until after lunch to get two rooms together. They went ahead and checked in, with the Baults taking Room 1404.

──◄►──

Hotel guests had little chance to be choosy in Atlanta in 1946. It was a seller's market. The Winecoff had been filled to capacity frequently since the war ended. Across America, new construction had been minimal for four years, as lumber was needed for the war effort and most potential buyers were overseas anyway. Returning GIs were now flooding the housing market in every city, often unable to find a place to live. In Atlanta, the large homes that lined Peachtree and West Peachtree going into downtown were filled with boarders. Those who couldn't find a room to rent or relative to move in with had to grab a hotel room, keep looking, and hope something would open up.

The housing shortage was deepened by the city's growth during the war. The metropolitan population had jumped from 472,000 in 1942 to 600,000, and the number of factories had increased by almost twenty percent in just the first fifteen months after the war. World War II had finally snapped Dixie out of the Depression, just as the federal government intended when it placed military bases throughout the region to supply needed infusion of capital.

The Winecoff was prospering in the new South, ranking among the city's top half-dozen hotels. The thirteen-story, "fireproof" Henry Grady, known for its backroom political deal-making, was a block north on Peachtree. The eleven-story, triangular Piedmont was a block to the south. Around the corner, two blocks down Forsyth, was the Ansley, with dining and dancing on its famous Rainbow roof. The 400-foot-long Biltmore, at West Peachtree and Fifth Street, and the Georgian Terrace, at Peachtree and Ponce de Leon, across from the Fox Theatre, were north of downtown.

Unlike the others, the narrow Winecoff did not have space for convention facilities or live entertainment, so it had to compete on service and location. It billed itself as "Nearer than anything to everything on Atlanta's most famous thoroughfare."

⋯ ▦ ⋯

When Pauline Bault realized the hotel was on fire, she was thankful that she had left her eleven-month-old baby home with the family's maid. As the Baults looked out the window, they could see heads hanging out of windows from the third floor to the tenth. Right next to them in 1402, a white-haired woman was crying out for help. Matsey Dekle had come to Atlanta from Tampa, Florida, with her daughter and son-in-law, Frances and Charles Pittmann. The Pittmans had gone on to Gatlinburg, Tennessee, to look at the lot they had bought for a summer home, leaving Matsey and her friend, Tona Perry, in Atlanta to shop.

The Baults watched as two aerial ladders on Peachtree began picking guests off window ledges, but they could tell the trucks could not move quickly enough. The firemen had started bringing out their nets. The Baults could see that a woman several floors below them was standing up on the window ledge, apparently ready to jump.

Sammie Sue Leadbetter had been by herself in 704 since just before midnight. That's when her husband, Tod, had gone to work as the overnight dispatcher at Plantation Pipeline's tenth-floor office in the Healey Building four blocks down Forsyth Street. He had been out of town all week and had called Sammie Sue and told her to pack a suitcase and meet him at the Winecoff. He said he wanted to see her but didn't have time to run out to their home in Decatur. Plantation Pipeline rented 704 permanently, and the Leadbetters had stayed there several times when it was available.

So for the previous four hours Sammie Sue had been alone and wide awake in 704. An habitual night owl, she hadn't even tried to go to sleep. She had thought about going down to see Tod at work but was afraid he'd scold her for being on the street so late.

Sammie Sue, a twenty-seven-year-old customer service representative in the telephone company's downtown office, had washed and rolled her hair and was standing nude, brushing her teeth, when she smelled smoke coming out of the vent. She took one step out of the bathroom and opened the door. The hall was black with smoke. After slamming the door, she thought about the man in 702. She had heard him earlier through the door that joined their two rooms and made them, when necessary, into a suite. Initially, she heard three men in the room. She could tell two of them were younger and the other was older. They were talking about getting some girls and had called the bellboy. The younger men finally said they would go find some girls themselves. They never came back. The older man — the others called him Bob — chose to stay in the room. Sammie Sue heard him call home and explain that he had a dental appointment the next day.

Bob Fluker was a forty-one-year-old safety engineer for J. A. Jones Construction Company. He had spent the past three years working on the Oak Ridge nuclear facility, but with that job completed, he was headed to Okinawa to help with the reconstruction of the air base there. Fluker had grown up in Gold Mine, Georgia, west of Thomson, where

his father had been superintendent of the mine. After graduating from
Georgia Tech, Fluker took a job with Tennessee Copper and then with
Homestate Mining in South Dakota. During the war he served in the
explosion control division of the Bureau of Mines.

Fluker's orders from J. A. Jones were to leave from Knoxville for
Sausalito, California, but because he would be overseas for two years, he
had decided to go home to see his parents in Thomson. He asked his
two nephews to come pick him up in Oak Ridge, and he made a
Saturday morning appointment with his dentist, whose office was near
the Winecoff. Because his nephews were going back to Thomson Friday
night, Fluker planned to catch the bus home, then return to Atlanta
Thursday to embark for Sausalito.

With her hair in curlers and cream on her face, Sammie Sue opened
the door between the two rooms and found Fluker still asleep. "The
hotel's on fire," she told him.

Startled at being suddenly awakened, he cursed her.

Retreating into her own room, Sammie Sue pulled on her camel's
hair coat, then opened the window but couldn't budge the screen. She
picked up a chair and cut a jagged hole with it. Fluker came into her
room as soon as he realized the hotel was indeed on fire. As they leaned
out the window, he put his arm around her and said, "I'm certainly
sorry."

They saw the fire trucks arriving and believed they would be rescued soon. So Sammie Sue dressed, packed the suitcases, and even
combed her hair. But smoke kept seeping through cracks around the
door and transom, forcing them into Fluker's room.

Sammie Sue and Fluker looked out the window and saw two ladders, a shorter one and a longer one. The long one reached the seventh
floor but seemed to be hardly moving. And as guests realized the ladders
would not reach them, desperation set in.

"Flames! We're burning!" yelled a man just above them.

Turning into the smoke, Sammie Sue made her way into her room
and to the hall door. She heard flames crackling on the other side. She
felt the door; it was hot, extremely hot. She returned to Fluker's window
and didn't look back again, keeping her face pressed against a wet towel.

They both heard the door fall in. Only Fluker turned to look.

"We've got flames in here. Get us a ladder quick," he yelled at the firemen.

But no ladder came, and the heat behind them felt like it was cutting through their backs. Fluker remained composed but talked little, except to repeat that he was sorry. Sammie Sue's hair felt like it was being singed and the smoke was suffocating. She kicked off her shoes and stepped up into the window. She felt jagged, hot metal burning her ankles and assumed the curtain rods had melted in two and fallen to the windowsill. A petite five-feet-three, she bent down and stepped out onto the window ledge. The room was so hot she couldn't even hold on to the inside of the window. She squeezed the hard bricks outside instead. She still did not hear Fluker say anything, and she was not about to turn and look at the approaching flames, which were being pulled by the draft of the windows, sucking up the room's oxygen and leaving only carbon dioxide to breathe.

Suddenly, below her, Sammie Sue saw a round white net with a big red "bullseye" in the middle. But just as quickly, the wind stopped blowing and the net disappeared. A cloud of thick smoke hung between her and the ground.

"Now's the time," she thought.

She bent her knees and jumped upward, pushing herself away from the building. For an instant, the air felt wonderfully cool and fresh. She hit the twelve-foot-wide net squarely, landing on her upper back. Her knees came up into her chest so hard that they cracked her sternum. A fireman and a woman bystander picked her up, put her on a stretcher, and placed her in an ambulance. She would live. Fluker would not.

The flames that burned down their doors had entered the seventh-floor hall from the stairway, which had been transformed into the perfect flue. Only a week earlier — Friday, November 29 — J. C. "Red" Kerlin had walked down the same stairway. Kerlin was the city fire inspector for the northwest section of Atlanta, and as was his routine, he had taken an elevator to the sixteenth floor and worked his way down through the building.

He checked the two-and-a-half-gallon, soda-acid fire extinguishers that were mounted in the hall of each floor. They all looked fine. He also looked for flammable material stored in dangerous places or in a dangerous manner.

Kerlin knew the Winecoff had no sprinklers, no fire escapes, and no fire doors. But he also knew the hotel had met the city fire code when it was built, and that was all that mattered.

The Winecoff had not had to install fire escapes because of the smallness of its lot. The 1911 building code mandated fire escapes only if the building's ground area exceeded 5,000 square feet. The Winecoff was built on a lot measuring approximately 4,400 square feet. Even if the hotel had been thirty stories tall, fire escapes would not have been required.

With no fire escape and with the elevators rendered inoperable by the fire, the only means of escape within the building was the stairwell, and it was impassable. Twenty-two years earlier, in 1924, the city had enacted an ordinance requiring fire doors in stairwells, but, again, the Winecoff had to comply only with the building code in effect when it was constructed. Kerlin knew that the Winecoff would not have been able to comply anyway. The law mandated that such doors could not open into the hall, but fire doors at the Winecoff would not have been able to open any other way. The lot was so small that the T-shaped stairwell was wrapped around the elevator shafts in the middle of the building, leaving no room for a landing at each floor.

The Winecoff's narrow width intensified the fire's draft, and seeking oxygen, the flames escaped the stairway onto the eighth floor almost as quickly as they had on the seventh floor. Flames were spreading down the front hall, where there had been two parties only hours earlier.

Navy flight officers had rented 806 and 830, diagonally across the hall, to celebrate their discharge from the service, and Dr. Renfroe Baird had rented 804 to party with several fellow interns and student nurses from Grady Hospital. But Baird had not drunk enough to impair his judgment when the fire awoke him less than two hours after he went to sleep. He had often thought about what he would do in a hotel fire, and now he was carrying out his plans — quickly. He stuffed a pillow in the transom, pulled the mattress up against the door, and threw the sheets, bedspread, and even the curtains into the bathtub. He was now making a rope of sheets with the help of the two student nurses who had shared the double bed with him. One of them, the former Elsie Newman of Clearwater Beach, Florida, was his wife. The other one — Helen Rutledge of Gastonia, North Carolina — was Elsie's roommate.

The two senior nursing students shared a deep secret: that Newman had been married to Baird since September. The Grady School of Nursing — believing that distractions were to be avoided and that husbands were distractions — forbade marriage. Newman had promised to sleep between her husband and her roommate.

Baird, twenty-four, had pawned a Winchester .410 shotgun to pay for the room and party. As a second-year surgery intern, he earned only ten dollars a month, plus room and board and a uniform, but he and the other interns liked to entertain. Baird and several others had pooled their money to rent a cottage on the Chattahoochee River north of the city. They called it Bedside Manor.

The son of a sessions judge in Pineville, Kentucky, "Sonny" Baird had come to Grady fresh out of medical school at Vanderbilt, where he had played football for a young assistant coach named Bear Bryant and had once scored three touchdowns in a game against Princeton. After eight months of medical school, Baird had become depressed and dropped out. He wanted to go to war, where his brother and all his friends were, so he joined up and was sent to Fort Oglethorpe, Georgia. Within a month the Army ordered him back to medical school. He was told he would be of more use to his country as a doctor.

The sheet rope that Baird, Newman, and Rutledge had tied together was more than twenty feet long. Sitting on the floor, with his feet braced against the wall, Baird pulled on the sheets as hard as he could, trying to yank his square knots apart. They held firm, but by this point Newman, only nineteen, was becoming so hysterical that Baird didn't trust her to be able to hold on to the rope. He told her she was going down on his back, with her arms wrapped around his neck. At six-feet-two and 200 pounds, he knew he could handle the load.

Baird padded the mullion between the windows with a pillow and then tied the sheet rope around it. Out the window they saw a virtual traffic jam of fire trucks, a stark contrast to the quiet street Newman had looked down upon when she awakened. The only person they had seen then was a man in an overcoat and hat across the street. Baird thought he looked like a bum. He had his hand cupped around his mouth and was yelling, "Hotel's on fire! Hotel's on fire!"

Baird could see an aerial ladder plucking guests from their windows, but he wondered if it would get up to the eighth floor in time, if it was even long enough to reach the eighth floor. He began stripping to his shorts, having learned in medical school that burns often occurred when steam was trapped inside of clothes and the clothes then caught on fire. He told Newman and Rutledge to take off everything except their bra and panties. He had already thrown Newman's fur coat out the window.

Baird and the two young women had tried to wet everything they possibly could so the room wouldn't explode in flames when the fire's gases were ignited by its heat. They had stopped up not only the tub but also the toilet and sink so they would have as much water as possible. But the water standing on the floor near the door soon began to boil, and when the mattress began to smolder, Baird said, "We're getting out of here." The ladder had just started moving toward their end of the building.

With Newman on his back, Baird started down the rope. He told Rutledge to come on behind. He felt responsible for the young woman and had tested the sheets enough to know they would hold. J. P. Sorrow, the driver of Ladder Truck 1, caught sight of them and quickly moved his aerial ladder to the window ledge of 604. One by one, the three climbed down the ladder. Grady Hospital's new red Cadillac ambulance was waiting for them on Peachtree Street, dispatched by an intern who knew they were there. They wrapped themselves in blankets and sped off to Grady to go to work.

It had been an incredible sight to the gathering crowd on Peachtree Street: three people dangling perilously on a white strand of cloth, escaping death by lowering themselves slowly through the smoke. Many guests above the eighth floor had also seen the miraculous escape and, at that moment, realized that they would have to do the same thing to live.

Associated Press

Having made the sheet ropes that they would later have to use, guests on the Peachtree Street side of the hotel look longingly for help. This photograph was taken before the explosive-like eruption of the flashover fire, which is pictured on the cover. (It was taken by Georgia Tech student Arnold Hardy, who won a Pulitzer Prize for the inset photo on the cover.)

The firemen were doing all they could, but their ladders simply weren't long enough.

The initial alarm had gone out at 3:42. Fire Chief Charles Styron had skipped the second alarm and sent out the third at 3:44. The fourth alarm went out at 3:49. The dreaded general alarm, summoning all off-duty firemen, was sent at 4:02: "Everything to Peachtree and Ellis, Winecoff Hotel."

The fire department had three main goals: to follow the flames up the stairs and through the hallways, to bathe the outside of building in water, and to rescue guests.

Soon Styron was summoning help from fire departments outside the city: East Point, College Park, Decatur, Avondale Estates, Druid Hills, Marietta, Hapeville, Fort McPherson, the Naval Air Station, and the Conley depot. More than forty fire trucks would flock at the corner of Peachtree and Ellis: thirty-two pumping engines, four aerial ladder trucks, six service ladder trucks, a floodlight truck, and a salvage and rescue truck. Against a high-rise fire in a building without fire escapes, the aerial ladder trucks would become the critical equipment in saving lives. The city had five, but more would be needed.

Chief Styron had been trying to order an additional eighty-five-foot ladder truck for several years, but none had been available with factories having retooled for the war. He had recently submitted a budget asking for $33,000 to buy an eighty-five-foot ladder and $44,000 to buy two sixty-five-foot ladders. Styron believed that any ladder more than eighty-five feet long was unsafe; he knew two longer ones had collapsed under their loads in other cities.

One of the fire department's existing eighty-five-foot ladders was sitting in the shop at Oak and Whitehall streets — a casualty of the sixteen-day nationwide coal strike. The truck, which belonged to Station 4 on Pryor Street, two blocks south of the Winecoff, had gone out of service on Tuesday. The leather cup that was part of its hydraulic raising device had worn out, and a new one had been ordered via air express from the manufacturer in Kenosha, Wisconsin. Normally, the fire department would have gotten overnight delivery from Kenosha, but the manufacturer warned there might be a delay because of the shipping embargo resulting from the strike. The part finally arrived Friday afternoon, and a repair crew started installing it immediately. Work stopped just before midnight and was to be completed Saturday. Extra mechanics could have been called in to work all night, but they weren't.

During the fire, the city's longest aerial ladder truck — Station 8's one-hundred-footer on Ellis Street — went out of service when the hydraulic system broke. Station 11's eighty-five-foot ladder quickly moved onto Ellis Street, leaving Station 1's seventy-five-footer and Station 5's fifty-five-foot mini-aerial on Peachtree Street. Their major obstacle was the fact that the Winecoff was 155 feet tall.

On the front side, guests above the eighth floor were out of the reach of the aerial ladders and left with three choices for survival: make a sheet rope to get down to the ladders, hope to wait out the fire, or jump toward a fireman's net.

The canvas nets, however, like the too-short ladders, were not designed to rescue people from the upper floors of a high-rise building. Styron told his men to bring the nets out when the ladders were extended nearly to their limit. He thought the nets, as a final option, might save another life or two. They instead became an enticement to death.

Firemen and spectators held the nets shoulder-high and, at the outset at least, beckoned guests to jump. And several guests on lower floors did so successfully. The firemen soon realized, though, that the nets were not designed for longer jumps. Some ripped on impact.

There were other problems. A net could lure more than one jumper at a time, and, on the front of the building, the marquee was in the way. One woman aiming for a net impaled herself on steel cable jutting from the marquee. Even without the obstacles, the net was difficult to hit from eighty feet up. One jumper who barely missed ripped the jacket off the back of a fireman in a final grasp at life.

In Room 902, the Meyer sisters had seen Sammie Sue Leadbetter's miraculous jump from 702. Both in their late fifties, the sisters were bookkeepers for Southern Coal and Coke Company in Knoxville. Their friend, Irene Wilson, a co-worker who had accompanied them on their Christmas shopping trip, was on the back side of the twelfth floor.

The Meyers lived with their mother and another unmarried sister in the family home built by their father, a German immigrant who had owned a Knoxville brewery until he was put out of business by prohibition. Their little brother, Billy, was the top manager in the New York Yankees' farm system.

Carrie, fifty-nine, climbed onto the window ledge first. "Hold my purse," she told Eva, "I think we're going to have to spend the night at another hotel."

When she saw a net below her, she pushed herself off the ledge with her hands and dropped. She hit the net, but the impact broke her arms, legs, and back. Eva, fifty-seven, balled herself up when she pushed off and broke only her arms and back.

No one had ever jumped into a net from a greater height and lived.

Bob and Billie Cox saw the Meyers' incredible leaps. In the room directly above the Meyers, Delilah Chambers had picked Little Bob up out of his baby bed, dressed him, and taken him into his parents' adjoining room — 1004.

Cox was trying to remain calm. He shone his flashlight up toward the Baults and waved, as if to say everything was OK. The Baults, still gasping for breath at the windows of 1404, were watching an unbelievable, fast-paced drama unfold below them. Rivers of water ran down Peachtree Street, through the curling hoses that crisscrossed themselves in a maze. Pumpers worked at every hydrant within three blocks, boosting the water pressure to the hoses that were assaulting the front of the building. Ambulances were descending on the Winecoff from everywhere, even the Naval Air Station and Fort McPherson.

The scene was like something out of a movie, and people were pouring in from all directions to watch. The first wave of spectators had been either travelers who had heard the sirens and walked up from the bus station three blocks away or guests in nearby hotels. But now residents were hearing the news and hurrying downtown from all sections of the city.

They watched as firemen Rick Roberts and David White alternately raced up and down Station 1's aerial ladders and driver J. P. Sorrow made life-and-death decisions. Firemen and policemen motioned frantically at guests stranded in windows. Some were screaming, others mutely pleading, and still others frozen in speechless terror, their beseeching faces back-lit by orange flames yet half-obscured by the smoke hugging the brown bricks. Any initial morbid fascination by the spectators ended with the mournful, piercing wail that traced the descent of the falling. The realization that people were dying before their very eyes sent many bystanders home quickly.

Billie Cox had told Delilah Chambers that if anything happened to them, she wanted Little Bob to live with Dr. Cox's family. Initially, Cox had told both women not to panic, that the firemen would rescue them. But as they stood at the window and watched the pandemonium below, it became increasingly obvious that they would have to use the sheet

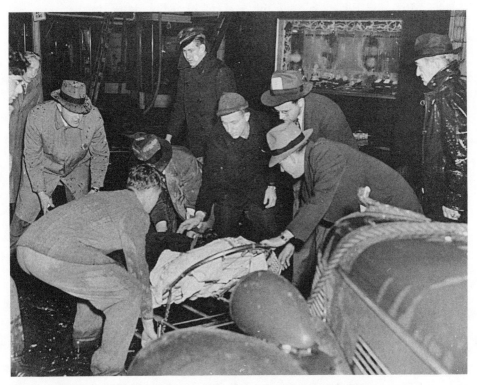

Civilians hurry toward an ambulance with a guest who fell to the sidewalk.

rope Cox had made.

Cox told Chambers she would go first; he felt responsible for her. As she backed out the window, he held her by the neck of her robe. She could hear two men talking in the window above her:

"Lord, what will we do?" one asked.

"All I know to do is pray."

Chambers prayed, too. She thought she was a goner. She felt the knots of the sheet tighten as her body turned in the wind. But within seconds she realized that she could make it. She felt so strong in her arms that she knew she could hold on until she reached the ladder. She had grown up on a farm and worked at a dairy for a decade, carrying jugs and milking cows. She was going down the sheet so fast that she believed she could go all the way to the sidewalk if the rope were long enough. The sensation of her feet hitting the ladder was like nothing she had experienced in her previous thirty-five years.

Now, it was Billie's turn. She had started crying and screaming when Chambers went out the window. Just three weeks earlier, Billie had suffered a miscarriage — a hard blow mentally and physically. She was still weak, too weak to hold on to the sheets.

At the top of the ladder Chambers felt something brush her arm. She glanced down and saw Billie hit the street.

Having watched his wife fall and believing her dead, Cox stepped up into the window clutching his young son. As he pushed off the ledge, Cox pitched Little Bob to the left, toward the hotel's awning. The men holding the net didn't see Cox coming through the smoke. His body hit the net, but his head struck the cast aluminum rim. Little Bob landed on the awning, bounced upward, and was caught by a man on the street. Cox would die an hour later, but his final act had saved his son's life.

Bob Bault saw it all happen. He also saw three other couples on the tenth floor try to save themselves on ropes of bed linen.

When flames entered 1006 — next door to the Coxes' suite — Louis and Jesse "Jake" Cochran had their sheet rope ready. But Jake lost her grip and fell through the awning and onto the dining terrace atop of the Winecoff's marquee. She broke both legs and arms and sprained her back.

The only way Cochran could get to Jake was to go down the same rope she had just fallen from. A forty-seven-year-old Miami car sales-man, Cochran had been in Atlanta all week on business. Cochran's daughter had invited him and Jake to stay at her apartment in Little Five Points, but they preferred to be downtown where they could shop. He had chosen the Winecoff because Allen Chappell, a friend of his and a member of the state public service commission, lived there. Cochran managed to drop down a story and half before flames leaping from the windows of 806 burned the sheets in two. His body hit a ladder — the same one a fireman had used to retrieve his wife from the awning — and bounced to the concrete.

A flashover had occurred, and flames were suddenly popping from the windows of the middle floors on all sides of the Winecoff. As flames had climbed the stairway — with the adjacent elevator shafts, it formed

Firemen dislodge a body — believed to be Jesse Cochran — from the dining terrace above the Winecoff's marquee.

a huge chimney — the fire's poisonous gases had risen to the top floor. Unable to escape through the Winecoff's unvented roof, the gases poured back down the stairway, meeting the rising flames on the middle floors, igniting, and exploding in a bomb-like flashover.

George and Edith Burch — the couple staying next to the Cochrans in 1008 — had married in April after meeting in Oak Ridge, Tennessee, and come to Atlanta for a sort of second honeymoon. They had just moved to Chattanooga, where the twenty-one-year-old Burch was in photography school. A native of Radford, Virginia, he had spent over two years with the 4th Marine Division, fighting in the Marshall Islands and Marianas. He was wounded in the head on Saipan.

Burch made a sheet rope and tied it to the radiator under the windows. He didn't use it until the door caught fire and flames ran across the rug and caught the curtains on fire. He yanked them down, and Edith started out the window. She tried to plant her foot in a small decorative inset in the bricks just below their windows, but her foot slipped. Then, suddenly, she no longer had the rope. It all happened so quickly she didn't even scream.

Burch watched his twenty-year-old wife plummet toward Peachtree Street. He saw her body break through the hotel's canopy, crash onto the dining area above the marquee, then roll off and drop to the sidewalk. Edith had broken both legs and arms and suffered internal injuries, but she was still alive. The canopy had slowed her fall.

Burch thought Edith was dead and knew he had to get to her. He nearly slipped on the window ledge and then somehow kicked his shoes off. He came down the sheet rope to the eighth floor, got on a ladder, and quickly climbed down to the street. But no one could tell him anything about Edith. He couldn't believe it. All the people standing there watching, yet none of them could tell him anything. Someone finally told him to look for his wife at Grady. He caught a ride in an ambulance.

The tenth-floor corner suite on the other side of the Burches — with its view of Stone Mountain — had been the residence of the same man for more than three decades: builder William Fleming Winecoff. He had named the hotel after his family, and, ironically, he and his wife,

Grace, would be among the 119 victims that would make the name a synonym for horror and tragedy. He died in a hallway and she on a sidewalk, after falling from a sheet rope. Both were seventy-six.

Winecoff had been born in Orlando, Florida, grew up in Brunswick, Georgia, and moved to Atlanta in the late 1800s. He arrived as a produce broker, but in a city that had been on the build since being burned during the Civil War, he quickly became a developer. He was one of the developers of prestigious Ansley Park, where his own granite and marble mansion had burned in 1910. Shortly thereafter, he chose the William A. Fuller Company of New York to design a fifteen-story hotel to be built on one of the highest points in the city. Architect W. L. Stoddard's plans called for a monument to concrete and steel, the marvelous new building materials of the day. The cost would be $350,000.

The steel frame went up first. Then floors of concrete were laid on a tile filler. The walls between the rooms were terracotta tile, plastered on each side. The brown-brick, Beaux Arts-inspired building was crowned by a white concrete facing that covered the bricks on the upper two floors. Stoddard was so impressed with his fireproof structure that he did not include fire escapes, sprinklers, or an alarm system in his design.

Winecoff operated the hotel for only two years and then leased it to the Robert Meyer chain of Birmingham, a well-respected operator of hotels in ten southern cities. The agreement called for the Winecoffs to be able to live there rent-free. They stayed even after the Depression forced them to sell the hotel to the Hightower family of Thomaston, Georgia, with the Meyer chain remaining as operators.

The Baults usually stayed at the Henry Grady when they came to Atlanta, and now they wished they had gotten a room there for Friday night. They had seen the flashover explode below them, slicing sheet ropes in two. They wondered how much longer they would be able to hold out in their own room. The man next door in 1406 had already told Bob Bault several times, "Cap, I don't think you can stay over there."

Bond Phillips — who had checked into 1406 with his wife about midnight — knew more about fires than anyone in the hotel. He worked in the automatic sprinkler construction department of the Grinnell Company. Twenty-six years old, Phillips had driven in from an out-of-town job in Chattanooga and rather than driving on to his home in Lithonia — about forty-five minutes east of Atlanta — had told his wife to meet him at the Winecoff.

Seeing little smoke coming from the Phillipses' windows and choking from the smoke in his own room, Bault finally yelled over to Phillips, "We're coming over there with you."

"How?" asked Phillips.

"Somehow," replied the Baults simultaneously.

Pauline Bault — thirty-six years old and 120 pounds — threw her coat toward the sidewalk, kicked off her shoes, and climbed tentatively onto the window ledge. The big searchlight immediately found her, and the people on Peachtree Street gasped helplessly. They couldn't yell and tell her to stay where she was, to wait for help, or how many people they had seen fall to their deaths.

Pauline turned her face toward the building and stretched her right hand toward Phillips. He grasped her hand, ready to pull her with one great yank into 1406.

Suddenly Pauline's body twisted around 180 degrees. The crowd below screamed in anticipated horror. Pauline was dangling below the window of 1406, only the strength in Phillips's right hand between her and death.

Phillips climbed up on the radiator in front of the windows. "I'm going to get you in here even if I have to break my back," he told her. He then heaved her up through the window.

Now Phillips had to worry about how he would get Bob Bault into 1406. He was a compact five-feet-eight and 150 pounds, but he knew that if he had barely been able to hold on to Pauline, then he certainly couldn't pull Bault into the room. He went to the bathroom and got a bedsheet out of the bathtub, knotted it to the radiator, and threw the other end to Bault, who tied it to the mullion between his windows.

Bault, a forty-two-year-old building contractor, had been paralyzed by watching his wife's dangling dance with death. For long seconds he stood transfixed, then finally threw his Dobbs hat out and crawled onto the window ledge. Using the sheet, Phillips pulled Bault into the room.

The suite above the Winecoffs' — 1108-10-12 — was being shared by well-known livestock auctioneer Florence Baggett and his two first cousins, Sarah Miller and Catherine McLaughlin. Baggett was a large, friendly man who seemed never to meet a stranger. He had driven his new 1947 Ford up from his home in Metter on Thursday to see a doctor about his enlarged vocal cords. His cousins came up from Gordon on Friday to join him.

Sarah Miller sewed canvas bags for one of the numerous kaolin companies in middle Georgia. The thirty-four-year-old mother of two children could have seen the job as a form of compensation. Her husband had died in surgery in 1943 after getting a cinder lodged in his throat while running a twelve-car train from the mine to the Gordon Kaolin plant.

When it became apparent that the three of them would have to use the sheets they had tied together, Miller made Catherine McLaughlin, her twenty-six-year-old sister, go first. Wearing a three-piece suit over her nightgown, she climbed down the sheets to a ladder and then on down to the street.

Miller and Baggett watched her disappear into the smoke and heard her call to them to follow. Miller knew it was her turn, but she was much larger than her little sister and nearly hysterical. Baggett tried to calm her and helped her climb out the window. Seconds later the flashover occurred and burned the sheets in half. McLaughlin was on the ladder when her sister fell past her. Baggett, stranded with no way out, died in the room.

Next door in 1106, Wylie Rochelle, a traveling salesman from Marietta, fell from his sheet rope and broke his neck. The first member of his family to graduate from college, he had survived the Depression by moving all over the South, as one company after another collapsed under him. When his son was drafted at age eighteen in 1944, Rochelle, then forty-three, had tried to go in his place.

In 1102, the corner room, Ed Thomas and Walter Baker decided to wait out the fire. They frequently stayed at the Winecoff because the Rhodes-Haverty Building — home of the zone office of the Buick Motor Division of General Motors — was only a block away. Thomas, fifty-five, owned the Buick dealership in Asheville, North Carolina, as well as Asheville Truck and Equipment Company. Baker, forty-one, was his sales manager. The two men were in Atlanta to buy war-surplus trucks; they had to have something to sell. New cars had been difficult for even dealers to get after the 1942 model, as car manufacturers started turning out engines for tanks, jeeps, and Army trucks. Since the war ended and the factories had been retooled to produce automobiles once more, demand had been so great that new cars and trucks were still hard to obtain.

Like fellow eleventh-floor guests Sarah Miller, Wylie Rochelle, and Eloise Mason — a fifty-nine-year-old retired Atlanta stenographer staying in 1104 — Baker died on the sidewalk of Peachtree Street. Thomas stayed in the room and suffocated.

This was not the first time Atlantans had stood downtown and watched people die. In 1936, less than three blocks away, hundreds of people going to lunch had watched in horror as the Cable Piano Company building burned on Broad Street. Only three people died, but the images and sounds were indelible: the terrified faces of the workers trapped at the windows of the upper floors, their chilling screams. The fire's rapid spread was blamed on the fire department, which was accused of being slow to arrive on the scene. Afterward, the city mandated that all nearby fire-fighting equipment had to answer a downtown alarm.

Only eighteen months later, on a May night in 1938, the Terminal Hotel burned, killing thirty-four people, many of them railroad men scheduled to go to work within hours. At the time, it was one of the worst hotel fires in U.S. history. Flames erupted at the five-story, brick-and-wood structure at the corner of Spring and Mitchell streets, near Terminal Station, shortly after 3:00 a.m. — about the same time the Winecoff would erupt eight years later. The fire was apparently ignited by an explosion in the basement of the kitchen and spread quickly. When firemen arrived two minutes after the alarm was turned in, the hotel was already engulfed in flames. The roof collapsed early on, and life nets were useless because the pall of smoke prevented the firemen from seeing the jumpers until they hit the street.

A rush of investigations, condemnations, and quick ordinances followed. Atlanta even passed a building code with retroactive features in 1943. Under the code, the Winecoff would have had to install fire doors and fire escapes. But before any alterations could be made, the city attorney ruled that buildings constructed under previous codes could not be required to make alterations.

In Room 1406, Bob Bault was helping Bond Phillips fortify the room against the fire. After throwing the sheets and curtains into the bathtub, Phillips had tossed the highly flammable pillows out the window. He then dragged the mattress off the bed and stood it against the door to block the smoke seeping into the room through the doorway and transom. He ripped two holes in the mattress with his pocketknife and, using the room's water pitcher, began filling the mattress with water. Working together, the two men dumped pitcher after pitcher of water into the mattress. When the water went off, they dipped the pitcher in the bathtub. When the tub was empty, they used the two glasses in the room to dip water out of the toilet. When that water was gone, they took a towel, sopped up the remaining water in the toilet, and wrung it out into the mattress.

During his trips to the window for fresh air, Phillips had tried to encourage the young men directly below in 1206 — Robert and Eaves Muns of Augusta. The two brothers were plumbers who had come to Atlanta to buy materials for their employer, John F. Street Plumbing Company. They drove to Atlanta Friday afternoon and had planned to go back that evening. But they got into town late and were going to have to buy the materials Saturday morning. They didn't even have any luggage when they checked into one of the Winecoff's last rooms.

Eaves, twenty-five, tried to jump before being restrained by his older brother.

"Wait a few minutes," Robert told him. "Help will come or we'll figure something out. If you jump, you know you're dead."

When their door burned through, they dropped their sheet rope out the window. Just as they started down, the flashover occurred, and flames burst from the windows of 1006, the room that Louis and Jake Cochran had just evacuated. The sheets caught fire, and they swung quickly to other sheets. Everywhere they turned, there was fire. The flashover had lit up eleven of the twelve rooms on the front side of the ninth and tenth floors. The body of a young girl glanced off Eaves' shoulder, but he didn't lose his grip on the sheets. Robert had to battle a flaming curtain and was burned on his left ankle, hand, hip, and face. Somehow they made it to the top rung of an aerial ladder, which gently swayed under their weight.

The flashover also engulfed the suite next door to the Munses' room, occupied by Joe and Gladys Goodson and their two small children. The Goodsons were on their way to Florida and had stopped to spend the night. They had reservations elsewhere, but the hotel gave their room to someone else. Reservations were held by the "honor system" generally, but ten dollars could prompt a desk clerk to find a vacancy in a full hotel.

Joe Goodson had been in poor health for more than a year. A geologist for Pure Oil, he had been seriously injured in a wreck a year earlier while working on an oil well in southern Illinois. After recuperating at his parents' home in Ormond Beach, Florida, he went back to work part-time in Illinois, but then in September contracted pneumonia and battled it for nearly two months. His heart had been weakened, and hoping he might benefit from a warmer climate, his doctors advised him to go back to Florida for four or five months.

At first Gladys held both her four-year-old son, Joe Jr., and twenty-one-month-old Barbara out the window, trying to protect them from the smoke.

"Don't throw them," yelled the firemen, as they scrambled to put a new crew in place with a net.

Gladys had no intention of dropping them — until flames burst through the door. Then she knew she had no choice.

They were so small they fell like rag dolls. The new net crew had not been able to get in place quickly enough. Gladys stepped up onto the window ledge and, screaming wildly, jumped for a net. Her back hit the rim, and she died instantly. Joe followed her. Truck driver Jack Mills picked up Joe Jr. and found him still breathing. But Mills knew he wouldn't be much longer. He would die, and so would his mother, father, and sister.

Charlie and Millie Boschung, two doors down the hall in 1208, had also been on their way to Florida — for their honeymoon. They had married Sunday in Cullman, Alabama, and were in south Georgia when Millie started feeling car sick and they decided to turn around and spend their honeymoon in Atlanta. They wired for rooms at the Ansley, but when they arrived Wednesday evening, their reservations — again, like the Goodsons — had been given to someone else. They took a room at the Winecoff and planned to head back to Cullman Saturday morning.

The Boschungs were tying sheets together when their door opened. A man and woman stumbled in through the smoke: thirty-two-year-old Ralph Moody, who was staying next door in Suite 1210-12, and University of Alabama student Rose Harvey, who was staying on the back hall in 1228. The two had met in the hall while trying to get on the elevators, which had stopped working in the fire's initial minutes. Unable to breathe or see, they had tried to get back to Moody's room but entered the wrong door. They knew they had to stay where they were.

Rose Harvey was working an internship at Rich's. Moody, who had a wife and child back home in Arlington, Georgia, had come to Atlanta to buy sawmill equipment. He already operated a garage but was considering returning to the sawmill business. Billy Newberry, an Arlington farmer, had come with him to pick up a "black-market" Mercury. Like many Americans in 1946, Newberry had given a dealer extra money — about $500 —to get a new car.

After going into the hall when the fire broke out, Moody and Newberry became separated at the elevators. As Newberry struggled to make his way back to 1210, he thought the person following him through the smoke and holding on to his waist was Moody. Instead, it was Alabama student Ethel Stewart, Rose Harvey's roommate. At the window they looked to the right and saw Moody and the Boschungs tying sheets together.

"That'll never work, Ralph," yelled Newberry.

Newberry believed the firemen would put the flames out soon, but Ethel Stewart, who was also working a Rich's internship, didn't think so. She was starting to panic. She talked about using a sheet rope to slide down to the vertical Winecoff sign on the corner.

"Let's just stay here for right now," said Newberry. "They're going to increase the water pressure in a few minutes and put this thing out."

The Boschungs figured they had a fifty-fifty chance of survival using the sheet rope. They thought they had no chance if they stayed in 1208. Millie went first. Instead of having her climb down the rope, Charlie tied it under her arms so he could let her down slowly. The twenty-one-year-old ex-Marine, a veteran of both Saipan and Okinawa, didn't doubt his ability to hold her.

Millie was calm when she kissed Charlie and started out the window. She made it to a ladder outside Room 808 and was reaching for a fireman's hand when the body of Edith Burch came hurtling toward her and knocked her off the ladder. She landed on the marquee. Charlie saw the whole thing, but still managed to use the sheet rope and get down to the ladder.

Next door, Ethel Stewart was talking about trying to jump onto the roof of a car on Peachtree Street.

"I don't think you could make it," Newberry told her. "Wait until they get some nets over here."

When flames entered their room, she asked Newberry to push her out the window. She said she couldn't make herself jump. He refused, saying that they should first see if Moody could make it down the Boschungs' sheet rope.

Moody had let himself down one floor when, inexplicably, he seemed to let go, his arms and legs going straight out. His body crashed all the way through the marquee and hit the sidewalk.

"He must've had a heart attack," Newberry told Stewart. "I know he's strong enough to hold on."

The two of them climbed out onto the window ledge, and as Newberry watched Rose Harvey climb out the window of 1208, he realized he would have to risk his life on that same flimsy sheet rope. Flames were jumping out of the top of their windows. He told Stewart to come with him, but she wouldn't move. As if in slow motion, Newberry arose from his crouched position, turned his back to the night air, and hugged the Winecoff. Carefully he took three sidesteps. Once on the window ledge of 1208, he pulled up the rope of sheets, nearly knocking Rose Harvey to the street as she climbed on the ladder at the eighth floor. Newberry used the rope to climb down to the same ladder. Ethel Stewart, engaged to be married in 1947, died on the sidewalk.

Smoke was pouring in around Bond Phillips' water-soaked mattress. Phillips had already closed the windows to cut down on the draft that was pulling the smoke through the doorway. But Pauline Bault kept opening the windows. She wanted to let the firemen know that they were still alive.

"None of us are going to be living very long if you keep doing that," Phillips told her.

All three of them were coughing now, as the room's supply of oxygen was running out. To get down to the remaining breathable air, they knelt between the foot of the bed and the windows and covered their faces with wet towels. Pauline told her husband she wanted to jump out the window into a net.

"No, we have to stay here together," he said.

But the smoke kept getting worse. Two doors down in 1402, Matsey Dekle and Tona Perry — the two elderly Tampa women who had pleaded for help — had already suffocated.

Suddenly Dorothy Phillips announced she was going to jump.

"I'm not going to get burned alive," she shouted, getting up and opening a window. Her husband was right behind her, though, and slammed the window down across her back. He pulled her back into the room.

Once again, the Baults and Phillips knelt on the floor. They began taking off some of their clothes as the room grew hotter and hotter. They heard the mirror on the back of the door crack and knew the door was about to burn through. Both Baults were becoming increasingly panicky.

"Try to stay calm," Phillips kept telling them. "A few minutes could change everything for us."

"Let's pray," said Dorothy Phillips. So together they prayed.

CHAPTER 4

⋯ ≡◆≡ ⋯

They were ready long before the Smiths got there. The Williams family had been anticipating the trip to Atlanta all week. For Boisclair Williams, it was going to be an opportunity to go shopping with her sister and to buy Christmas gifts she couldn't dream of finding in Cordele or anywhere else in south Georgia. For her seventeen-year-old son, Ed Kiker Williams, it was an opportunity to skip school and, more important, to drive Aunt Dot's new blue Buick. For his eight-year-old sister, Claire, it was the day she had been dreaming of, the day she would see *Song of the South*. It would be months before the movie got to Cordele, so she was going to see it before any of her friends — in a city called Atlanta, a magical place that she had never seen but had heard so much about.

The Smiths made it from their home in Fitzgerald to Cordele by mid-morning Friday. Dr. Smith had stayed home to work, as Saturdays were a busy day for appointments. Ed Kiker's father, cotton warehouseman John Williams, was also staying home.

Still, with Aunt Dot and her three children — Fred, Dotsy, and Mary

— there were seven people in the Buick when Ed Kiker turned it north on Highway 41. It would be a four-hour trip over a two-lane highway bracketed by small towns and country hamlets. About an hour out of Atlanta, the city's hotels and restaurants began popping up on billboards. Everyone in the car noticed the one advertising the Winecoff, since that was where they had reservations. The sign described it as a "fireproof hotel." They hadn't realized that. They had chosen to stay at the Winecoff simply for convenience, but now they felt even better about their choice.

At check-in, Boisclair and Dot asked for rooms near each other. The Smiths needed a room with two double beds, so they were given 1530, the biggest room on the fifteenth floor. The Williamses took 1520 on the back hall and asked for a roll-away bed to be brought in.

The women decided to go on to the movie that afternoon — Claire and Dotsy couldn't wait any longer — and then shop after dark. The Disney masterpiece, taken from the Uncle Remus tales of Atlantan Joel Chandler Harris, was starting its fourth week at the Paramount. The families walked across Peachtree Street in time to catch the 3:25 show.

After dinner, the Smiths and Williamses shopped for a while at Davison's. As they returned to their rooms, they agreed to be up and ready to shop again when the stores opened on Saturday. They planned to spend the morning at the huge Rich's store on Forsyth Street.

Rich's and Davison's had drawn not only shoppers to the Winecoff but also workers. Rose Harvey and Ethel Stewart were just two of nine students from the University of Alabama staying at the Winecoff while working internships at the two department stores. All were home economics majors, specializing in clothing and textiles. As a requirement for graduation, they had to complete the three-week internship for which they would receive three hours of classroom credit. The internships were scheduled to provide sales help during the Christmas shopping season.

The students had arrived by train on Sunday and were to work until the day before Christmas Eve. Eight of the nine were seniors and had anticipated the trip throughout their college careers. They were through with exams for the quarter and looking forward to being in a large city at Christmas with no adult supervision, not even a housemother.

Like the Williamses, the students were staying on the Winecoff's back side — in some of the least appealing rooms. Instead of giving the young women rooms on the same floor so they could visit easily, the hotel had stacked them on top of each other at the end of the back hall, in 728, 828, 1028, and 1228. The rooms had single beds and were among the Winecoff's least expensive.

Gloria Riemann and Eloise Eicholz were in 728. Lifelong friends, they had grown up together in Gulfport, Mississippi, and attended Gulf Park College as freshmen before transferring to Alabama and joining Delta Gamma sorority.

On Friday evening, though, Eloise had gone back to Tuscaloosa to attend a fraternity formal Saturday night. She had originally planned to leave Saturday after work, but her boyfriend persuaded her to take Saturday off and come Friday night, when he knew he could borrow a car to meet her at the Birmingham airport. Uneasy about staying alone in a hotel room in a strange city, Gloria asked Elaine Sullivan to move in with her for the weekend. Elaine was rooming with Sarah Aldridge and Jean Pruett in 1028, where a roll-away had been brought in, so she was happy to have a little more space for a couple of days.

When the smoke in 728 awoke them, Gloria and Elaine went to the windows. They could hear the man below them talking to their classmate Peggy Cobb, perched in her nightgown in the window directly above them. Peggy was a native of Fayette, Alabama, and worked on the *Crimson and White*, the college newspaper. The man — Nelson Thatch, the Winecoff's day clerk — was pleading with her not to jump. He shouted for her to hold on until a fireman could get to her.

But when the fire burned down the door to 828, Peggy jumped. Her roommate — Fredna Green, a senior from Hattiesburg, Mississippi — remained on the ledge, wrapping her arms around the mullion between the windows and telling herself over and over, "I can't turn loose. I must hold on. I mustn't go to sleep." Finally overwhelmed by the smoke, she passed out. When a fireman rescued her from the window ledge, her hands were still squeezing the mullion.

Gloria and Elaine didn't jump either. They suffocated before flames consumed 728. So did Elaine's roommates in 1028, Sarah Aldridge and Jean Pruett.

⊶ ⩷⩸ ⊷

In 1520, Ed Kiker Williams awoke to his mother's voice.

"I don't know what that is," he heard her saying. "I think it's a smoke bomb."

She got up and switched on the overhead light, barely visible for the smoke. Ed Kiker jumped up and closed the transom, then ran to the windows and opened them.

The Williamses thought of their relatives, the Smiths, down the hall. They cracked the door to see if they could sprint to the Smiths' room, but slammed it shut when they saw they couldn't. Boisclair yelled as loud as she could through the closed door. She thought she heard her younger sister respond. "Stay in your room," screamed Boisclair.

By now the smoke was so thick in 1520 that the Williamses could not even see each other. They huddled together at the double windows. Gasping for air between the billows of smoke rushing into their faces from the lower windows, they talked about what to do.

"I think maybe I could jump over to that roof," said Ed Kiker. "Then I could get some help and come back and get you all out of here."

The roof of the Mortgage Guarantee Building was directly across from the fifteenth floor of the Winecoff. Only ten feet away, it looked close enough to reach out and touch.

"I don't think you could make it without a running start," said his mother.

"Probably not," he sighed.

Jack Sheriff had tried to jump the alley from his room on the seventh floor. He thought he could smash his hands through the windows of the Mortgage Guarantee Building, unaware that they were made of wired glass. He didn't make it anyway. His girlfriend, Margaret Nichols, didn't make it either. He had tried to throw her across the alley.

Sheriff, thirty-two, owned a taxi company in Washington, a laundry in Maryland, service stations in both places, and an Atlanta nightclub. Jack

Sheriff's Theatre Restaurant was next door to the Winecoff, above Chandler's Shoe Store. A Japanese restaurant — Wisteria Gardens — had closed there after the war started, and Sheriff had opened a cocktail lounge and dance hall.

Nichols was the club's press agent, cashier, and hostess. Nicknamed "Luscious," she was a striking, thirty-year-old redhead, a onetime Miss Atlanta runner-up, and well-known former box-office girl at the Fox and Paramount.

After closing Friday night, Sheriff had returned to the room he rented permanently at the Winecoff. Nichols changed out of her evening dress, and they then went back out to eat, returning to the hotel at 3:10, roughly thirty minutes before the fire broke out.

The smoke grew so thick in 720 that Sheriff convinced himself the only way they would escape was to jump across the alley. And if they didn't make it, he reasoned, then falling to their deaths would be preferable to burning to death. After Nichols fell to the alley below, Sheriff followed, missing the other building by inches.

Bill Bryson tried to jump to the Mortgage Guarantee Building as soon as he awoke and realized the hotel was on fire. He had been asleep in 928 — one of the back-corner rooms — with two other bus drivers for Smoky Mountain Trailways. Because drivers who made runs of more than four hours were required to have eight hours off before making a return trip, Smoky Mountain Trailways rented 928 permanently. The bus station was only three blocks away on Spring Street.

The drivers considered the Winecoff a first-class hotel, even if they did have cramped quarters. Two single beds and a roll-away were jammed into the room, and the men would come and go depending on their schedules. Olan Hart had returned to the station at eleven o'clock to make a midnight run, leaving both Bryson and Cleve Sisk asleep in the room. Harold Sorrells came in later.

The son of the dietician at Western Carolina Teachers College, Bryson was slender, redheaded, and outgoing. A wartime sailor, he had returned home to the mountains from the Pacific in 1945 and gone back to driving a bus. Two weeks earlier, he had moved his wife and two little girls from a trailer in Cashiers to a house in Asheville, bringing an end to his daily hundred-mile commute.

In November, Bryson's wife, Vera, had come to Atlanta with him and stayed at the Winecoff. While they were waiting on an elevator, she had spilled some ashes from a cigarette. "Don't worry," he had told her, "this is a fireproof hotel."

But the twenty-eight-year-old Bryson was actually terrified of fire and had even imagined his own death by burning. His parents' home had burned while he was in the Navy, and he had told his wife that there had been a destructive fire in every generation of his family. He had even told her he would die in a fire when he was twenty-eight, but he had also vowed that he would do anything before he would burn to death.

So as soon as Bryson realized how bad the fire was, he made his panicked leap for the Mortgage Guarantee Building. Sisk and Sorrells, both of whom had been driving buses through the Smoky Mountains for more

A burned-out room on the back side of one of the middle floors. The Mortgage Guarantee Building — only ten feet away out the window — was so close it invited several fatal attempts to leap across the alley.

than a quarter-century, did not follow him. Sisk died in the room, while Sorrells retreated to the window ledge and fell to the alley only after the fire seized the room.

The guests on the back side of the Winecoff were trapped. There were no aerial ladders, and only a few people on lower floors resorted to sheet ropes. Whereas guests on the front side were influenced by the power of suggestion — seeing others both escape and die via the makeshift ropes — those on the back could see nothing but gray-black darkness. Firemen had followed Jimmy Cahill's lead by placing boards and ladders across the alley, but they simply could not rescue the guests quickly enough. Some guests made wild leaps toward a ladder, hoping they could grab it in mid-flight and be saved. One of them was sixty-four-year-old E. B. Weatherly, a frequent Winecoff guest and one of the most prominent people sleeping in the hotel that night.

Weatherly was past thirty and a father several times over when he enrolled in law school at Mercer College in Macon. He went to class by night and sold typewriters by day.

A year after the Depression hit, he moved his family to his 4,500-acre cotton farm hard by the Ocmulgee River. It was Christmas, 1930. "A lawyer will starve to death in town," he told his wife and six children, foreshadowing the suicides of attorney friends who remained in Macon.

Weatherly's family had plenty to eat on the farm, and he became a tireless crusader for agricultural revolution in Georgia. With his well-known declaration that he had "$50,000 less than nothing in cotton," Weatherly was consumed with breaking the economic chains of a one-crop economy.

He decided to try livestock because it had some of the same government protection as steel and other manufactured goods. He raised Black Angus cows, Hampshire hogs, and Corriedale sheep and sold cattle for breeding all over the country. He became the premier cattleman in the South. By 1934, the U.S. Agricultural Adjustment Administration had appointed him to a beef industry committee that he would later chair.

Whenever Weatherly came to Atlanta to ask the General Assembly to increase appropriations for the cattle industry, he always stayed at the Winecoff. He had become good friends with the manager, Lasso Moseley, and continued to stay there even after Mosley was let go by the hotel's new operators in 1945. He always took a room on the tenth floor or above, so as not to be bothered by the noise of city traffic.

Weatherly had come to Atlanta Friday to see a highway department administrator. He wanted to discuss improving a road leading to a meat curing and packing plant he had built on his farm. He planned to produce top-quality hams and sausages and truck them to Macon and Atlanta for direct delivery to customers. But the man he needed to see at the highway department was not in Friday, and Weatherly was told to come back Saturday morning.

Weatherly had asked his wife, Bessie, to come with him, but she had decided to stay home and get an early start on Christmas. The children and grandchildren were coming home for the holidays in a couple of weeks, and she had already picked out a Christmas tree. She planned to have it sprayed white over the weekend.

Weatherly was in one of the "24" rooms — 924, 1024, etc. — all of which had blind-like shutters on the outside of the windows to afford privacy from the offices ten feet away. Because the Winecoff was not air conditioned, the drapes in the "24" rooms needed to remain open so the stairway could be ventilated through the transom. After finally breaking through the shutters, Weatherly stood up on the window ledge. He saw a ladder through the smoke below and made his leap for it, but he missed and his body fell to the alley below.

In 824, another shuttered room, Lena Harris and Irene Tollett tried to calm each other as they waited. They were sisters, daughters of the owner of a general store in Isabella, Tennessee, one of the three small towns in an area on the state's southeast border known as the Copper Basin. Unmarried, Harris still lived there; she worked as a cashier at the First National Bank in Copperhill during the week and taught Sunday school on the weekend. She came to Atlanta several times a year to visit her sister. They loved the theater and opera.

Atlanta was Tollett's favorite city, and she had finally moved there in 1943 to become the librarian at Russell High School in East Point. She and her husband, an attorney, divorced in 1931 when her son, Bobby, was only four. The Depression South was not a friendly place for divorced women, and Tollett had moved back home, where she became an English teacher and then a librarian. In the summer, she took classes at George Peabody College for Teachers in Nashville and ultimately earned a Ph.D. in library science in 1942. She then landed the job in Atlanta.

Tollett had beaten the odds by successfully supporting herself and her son. He was now a senior at the University of Georgia, and she was looking forward to his graduation — and an alleviation of the pressure she had lived with for the past fifteen years.

Harris had arrived Thursday and walked two blocks up the street to the Winecoff after being turned away at the Piedmont, where she always stayed. She had intended to take the train home Friday night — the route from Atlanta to Knoxville ran through the Copper Basin — but decided to stay and go to dinner and a movie with Tollett. They went across the street to the Loew's Grand to see Katharine Hepburn in *Undercurrent*.

Tollett had a meeting at school at eight o'clock Saturday morning. Because catching the trolley to the meeting would be quicker from downtown, she had decided — at her sister's invitation — to spend the night at the Winecoff rather than go home to her apartment at Peachtree and 15th Street. Entombed in their room by the shutters, the two sisters suffocated together.

―• ≡♦≡ •―

Once Ed Kiker Williams had given up on the idea of jumping the alley, Room 1520 became quiet. After several minutes, Boisclair asked her son what he was doing.

"I'm praying," he replied.

"Good," she said, "that's what you ought to be doing. Because nobody but God can save you right now."

Again, the talking stopped. Against a backdrop of wailing sirens and human screams, mother, son, and daughter gasped for air at the windows.

"Mama," Ed Kiker finally said.

There was no reply.

He turned around but could see neither his mother nor his sister, so he assumed they must have passed out and fallen to the floor. He took a deep breath and began to feel for their bodies. He found his mother's first and tried to lift her to the window, but he needed another breath. He lunged forward, leaning out the window and inhaling deeply, then returned to his mother. This time he got her up to the window and propped her body so that her head was on the window ledge. After inhaling deeply once more, he reached down and found his little sister. He picked her up and put her in the window, too, laying her body almost on top of his mother's. Exhausted and heartbroken, his eyes and throat stinging, Ed Kiker stuck his own head out the other window and panted for fresh air.

Ten feet away, next door in 1522, an hysterical Jane Wallace was ready to jump. Suddenly she heard someone banging on her door, trying to get in. When she opened it, Carlos Hamil and his young son stumbled in. Hamil was a math and science teacher at Boys High School in Rome and a chaperon for its Hi-Y delegation. His nine-year-old son Richard had awakened him in 1524.

"There's something bad wrong," Richard had told his father.

But Carlos had dozed back off — exhausted from the day's trip, two sessions of the Youth Assembly, and being responsible for eight high school boys. After the evening session, he and Richard had run into Mary Alice Davis, the chaperon from Bainbridge High School, on the elevator.

"I'm worn out," he told her. "We're going to bed."

"So am I," she replied.

Davis and her seven students were already struggling for their lives when Richard shook his father again. This time Carlos remembered his eight students. He and Richard headed into the hall, trying to get to the boys in 430 and 1030.

Other guests were in the hall, too, running and screaming. Carlos was looking for a fire escape, an elevator, stairs — anything that would take him and his son out of the hotel. Instead, they found only more smoke. By the time they realized they would have to go back to their room, they were so disoriented and blinded they went next door to 1522 instead.

"I'm in the wrong room," Carlos said when he saw Jane Wallace.

"It doesn't matter," she told him. "We're all going to die anyway."

Hurrying back to the windows, she told the two strangers, "I'm going to jump."

"No, you're not," Carlos shot back.

"Well, I'm not staying in here and get burned alive," she replied.

"I'm not either, but they might get the fire put out before it gets to us. There's no point in jumping now."

Carlos took Wallace by the arm and tried to reason with her.

"Let's not jump until we have to," he said. "Where are you from, anyway?"

"Holly Springs, Mississippi."

Wallace had come to Atlanta to talk to her husband, Byron, whom she had met during the war while she was a nursing student in Oxford, Mississippi, and he was in Army training at the University of Mississippi. After he left Ole Miss to go to war, she never saw him again.

She had caught the train to Atlanta Friday. Her husband lived in College Park, and she planned to make one last attempt at a reunion before divorcing him. She was twenty-four and ready to get on with her life. If Byron didn't want to be married to her, she was sure someone else would.

As the trio huddled by the window with towels over their faces, Hamil's mind kept going back to the eight boys he had brought to Atlanta. He wondered if the fire might not be as bad on the lower floors. Maybe it had started above their rooms. Maybe the boys had already gotten out of the hotel. Or maybe some of them were already dead. How would he tell their parents?

Jane Wallace and young Richard leaned out the windows for more air, but it didn't do much good. The space between the two buildings had become a huge chimney. Thinking the firemen would not be able to see them, Richard put a lamp in the window. It burned until the electricity went out.

By then even Hamil was beginning to doubt if they could survive. The smoke in their room was getting heavier. It was becoming impossible to get any fresh air, and the flames below them were still rising. Quietly, Carlos murmured to Richard, "I don't think we're going to make it."

Richard had not wanted come to Atlanta anyway. He didn't like missing school. But Carlos had thought it would be good for him to come to the city to see the lights of Christmas. Suddenly Richard understood why he was there: somebody had to encourage his father.

"No, Daddy, something's going to happen. We'll get out," replied the boy.

So there they waited, a man and his son and a woman they had never seen before, gasping for air and searching for help, side by side because on this one night of their lives they had been assigned adjacent hotel rooms.

Strangers had similarly been tossed together in Room 1018 on the hotel's back side. John and Dick Turk, two brothers not long back from the war in Europe, had been standing in the hall pounding on the door when Kelly and Dorothy Hammond finally let them in.

The Turks immediately started looking around the room for something that would stretch across the alley. The Mortgage Guarantee Building looked so close that they thought of jumping across. Dick picked up the room's metal water pitcher and flung it toward the window opposite theirs, shattering the glass. Nevertheless, the Turks' search turned up nothing that would stretch the ten feet across the alley. But unlike the other rooms on the back hall, the corner "18" rooms offered their guests another possible escape.

Looking out the two single windows above the Ellis Street side, the Turks could see Ladder Engine 11 parked on the sidewalk next to Davison's, creating a less steep incline than from the middle of the street. But the big aerial ladder was picking off guests one by one on the lower floors, and the Turks knew it would be a long time before it worked its way up to 1018. They wondered if the ladder could even reach that high.

Because the Hammonds had closed the transom before going to bed, the smoke in Room 1018 was tolerable. The four occupants talked of how they could escape. The Turks even had time to think of their brother Tom on the fourth floor. He was the reason they were both in the Winecoff that night.

John had brought Tom and his family to Atlanta Friday afternoon to catch the next morning's train back to Oklahoma, where Tom was stationed in the Navy. On a two-week leave, he had come home to Butler, Georgia, to visit his mother over Thanksgiving. Since the train to Tulsa was leaving early Saturday morning, the Turks decided to spend

Friday night in a downtown hotel. Tom, his wife, Orlena, and their two-year-old daughter, Nancy, got a room on the Peachtree side of the fourth floor. John was given one on the back side of the tenth floor.

Dick lived in Atlanta. When he got out of class at Georgia Evening College downtown around nine Friday night, he headed by the Winecoff to see his brothers. They talked for several hours in Tom's room, 406, laughing, reminiscing, just enjoying being together. Dick, twenty-eight, didn't really want to leave. He had only been home from Europe since April, and it was nice to be around his older brothers. Ever since Tom had moved to Oklahoma before the war, such family get-togethers had been few. Their father, a small-town doctor, had been dead for several years. During the war the brothers had often wondered if they all would ever be together again.

Dick finally said he had to go in order to catch the trolley out to his apartment on West Peachtree Street. But John persuaded him to stay; there was plenty of room in the big double bed.

John was a heavy sleeper. His little brother awoke first and punched him.

"Let's find the stairs," said Dick, pulling on his pants.

Room 1022 was only fifteen feet from the stairs, but the smoke blinded the two men and forced them to feel their way along the back hall. By the time they got inside 1018, Tom and his family were already sitting on a window ledge of 406, waiting to be rescued by the firemen who had just pulled up on Peachtree Street. They were the first people down the ladder; the firemen had reacted quickly when they saw little Nancy. Tom immediately called his mother.

In 1018, Dick was already imagining his mother sitting in a church with three caskets below the pulpit. "Oh Lord," he thought, "they're going to have a great big funeral back home."

The Turks paced from the windows above Ellis Street to the ones over the alley, alternately cursing and praying. Dorothy Hammond was talking about dying.

"No, I don't think it's our time," her husband told her repeatedly.

Kelly Hammond was a thirty-two-year-old regional representative for Allis-Chalmers, the big midwest tractor company. He covered an

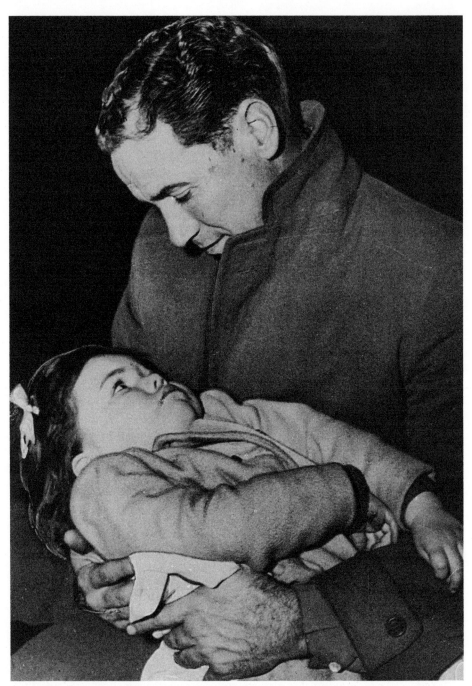

Tom Turk embraced his daughter Nancy on Peachtree Street after being rescued from Room 406. His two brothers were still trapped on the tenth floor at the time.

area of middle Georgia stretching from Eatonton to Americus and had come to Atlanta for a sales meeting. Dorothy had come along to shop. As usual, they had reserved a room at the Winecoff, near Allis-Chalmers' regional headquarters on Spring Street. Kelly had first stayed there in 1940, after seeing one of the "absolutely fireproof" ads. They planned to check out Saturday morning.

The smoke became so thick in 1018 that the Turks and Hammonds resorted to breathing through wet towels wrapped around their heads. When the electricity went off, they decided the aerial ladder on Ellis Street would most likely move to windows where the firemen could see the most people. They all clustered in one window.

Meanwhile, the battle to quell the flames waged on below. One floor after another was being claimed, but the advance was slow and grueling. Over and over the firemen repeated the same procedure: go down the steps to the trucks, grab a fifty-foot section of rolled-up hose, carry it up the Winecoff's labyrinth of stairs, clamp the active hose line shut, remove the nozzle, add the new section of hose, lay it out, affix the nozzle to the end of the dry section, signal for the removal of the clamps, and then hold on for dear life and fight the fire until the next section arrived.

With the furnace-like heat blistering their faces, black smoke choking their lungs, and hot water pouring down on them, the firemen felt as if they had gone into hell. Many of them had been back from the war less than a year; they wondered if they would survive again or had instead come home to die.

At about 4:45, the firemen working on Ellis Street shone a spotlight on that side of the building. The Turks and Hammonds now knew the firemen realized they were there. The aerial ladder had reached the ninth floor, but they could tell that was as far as it could extend. They knew they would have to get down to it somehow, so they began tying blankets together.

Meanwhile, directly below them in 918 LiLa Johnston was having difficulty getting on the ladder. She was a large, fifty-seven-year-old woman, and although the ladder was as high as her window, it didn't quite touch the building. There was almost a foot between the window ledge and the top rung. LiLa — who had come to Atlanta from Eufaula, Alabama, to care for her daughter at St. Joseph's Infirmary only three blocks away — was trying to come out the window head-first. The fireman at the top of the ladder told her to back up into the bathroom. He followed her through the window, then brought her back out with him.

Realizing how close they were to rescue, the Turks and Hammonds knew they had to hold on another few minutes, that they could not panic. In 818, the room directly below LiLa Johnston's, an exhausted James Little had been rescued on a ladder stretched from the Mortgage Guarantee Building, but his good friend, Frank Jones, was already dead.

Both men worked in the Manhattan Project at Oak Ridge, Tennessee, the site of the research that helped develop the two atomic bombs dropped on Japan sixteen months earlier. They left Oak Ridge after getting off work Friday evening. Jones, a twenty-one-year-old clerk-typist, had left a note in his dormitory mailbox: "Have gone to Atlanta. Be back Sunday afternoon."

For forty-five minutes in Room 818, Little physically restrained Jones to keep him from jumping. Their opposite reactions reflected the range of response of guests throughout the Winecoff. Some — despite being challenged physically and emotionally — were able to remain calm, allowing their minds to think clearly and their adrenaline to summon all their body's strength. Others were overwhelmed by panic. Jones finally broke away from Little and jumped to his death, only five minutes before the ladder appeared through the smoke.

The Turks and Hammonds had maintained their composure. By the time LiLa Johnston had climbed down the ladder, they had tested their blankets by pulling on them as hard as they could. The Turks told Dorothy to go first. She refused, insisting that John go first, that she wanted to go with her husband.

"That way, if something happens, we'll die together," she said.

John agreed to try the rope of blankets first. The fireman on the ladder was going to get a coil of rope, but John wasn't waiting on him. He was barefooted, and the floor was getting hot. He knew they didn't have much more time.

With one end of the blankets tied to the radiator, John took the other end and backed out the window. Dick held him under the armpits until he could lower him no farther. Dick then held the blankets and let them out slowly. The smoke outside the window prevented him from seeing if John made it to the ladder, but when there was no longer any weight at the other end of the blanket, he realized his big brother had either made it or fallen to his death.

As Dick pulled the blankets back up, a ball of coiled Manila rope flew into the room. The fireman had given it to John, who threw it through the window from the top of the ladder. Dick and Kelly tied the rope to the radiator and then let Dot down to the ladder. Kelly quickly followed.

Dick, still shirtless, climbed out the window last. Flames were leaping from windows on the eighth floor, but he needed to get only to the ladder at the ninth. The sensation of his bare feet hitting the top rung was the best feeling of his life.

The fireman at the top of the ladder heard someone holler, looked to his left, and saw Gregory Vojae dangling from a sheet rope below 1014, two rooms away. Vojae had prayed fervently to remain calm, climbing onto the window ledge only when the heat had blistered the paint on his door and fleeing on his sheet rope only when flames entered the room. He knew he had only minutes before the flames marched across the floor to his rope.

The fireman swung the Manila rope toward Vojae, but Vojae was clinging to his sheet with his right hand and missed the rope with his left. Again the fireman swung the rope, and this time Vojae caught it. With the Manila rope doubled around his fist, Vojae at last let go of his sheet and swung slowly toward the ladder and the arms of the fireman.

⊶ ⛉⊹⛉ ⊷

In 1520, loneliness was overwhelming Ed Kiker Williams, and his hope for survival was all but gone. He could see no one through the smoke, and he kept thinking about his mother and sister. He had done everything he could for them, but he was beginning to believe they were already dead. If they weren't, he assumed they would be by the time he was rescued. He figured they might have already gone to heaven.

The thought of losing his mother was more than Ed Kiker could bear; he really didn't want to go on living without her. Still, something kept telling him he would have to live without her, that somehow he would survive. As the smoke in his lungs increased, Ed Kiker leaned farther and farther out the window. Finally, he had to wrap both his hands around the mullion between the windows to keep from falling out. But the more he leaned out, the more the smoke seemed to engulf him. He drifted slowly into unconsciousness.

The sudden sensation of falling jolted Ed Kiker back to awareness. Somehow he had managed to hold on to the mullion, but now he was dangling below the window, his arms already aching from the strain. He knew he would eventually lose his grasp.

Three floors below, Nell McDuffie was backing slowly across a ladder to the Mortgage Guarantee Building. She was assistant manager of the Winecoff, having worked there for seventeen years. She did secretarial work for the Geeles, the hotel's operators, and helped out on the front desk — whatever was needed.

When the ladder was moved to 1120, fireman William Holt had asked Nell if she could get on it. She hollered to Holt that she thought she could make it. The ladder was hooked firmly to her windowsill, but as she began to back down, her foot slipped. Holt again called to her to ask if she needed help. Again she said no.

All the way across the ladder she could hear the calming voice of Holt, a fireman who worked at nearby Station 8, on the corner of Carnegie Way and Spring Street. Now, in the middle of a chilly, desperate night, his voice was a lifeline to her, mesmerizing as it came through the smoke and whining sirens. He talked her all the way across the ladder.

"Take it easy, honey," he called out. "You've got plenty of time."

When Nell made it to the other side, she climbed down into an architect's office. The wind had whipped at her on the ladder, and she was cold and wet and shivering. Fireman Red Pittman gave her his jacket.

Suddenly they all heard a noise. They turned to look out the window, and there on the ladder lay Ed Kiker Williams. The smoke had finally overcome him, and he had passed out and fallen from his window. He landed on the ladder — twenty-five feet below — that Nell McDuffie had just climbed off of. His body had bounced up slightly upon impact and then landed on it again. Jolted back into consciousness, Ed Kiker heard a voice: "A boy just landed on the ladder."

Then, louder, he heard someone call to him: "Can you climb off there, son?"

Ed Kiker raised his head and said he could. But just to make sure, fireman "Big Boy" Anglin, all 250 pounds of him, climbed onto the ladder to help him.

Ed Kiker's head was cut and bleeding where he had hit the ladder, and he felt a sharp pain in his foot, but all he wanted to do was tell the firemen that he thought his mother and sister were dead. He was crying and begging them to go to the roof, crawl over to the fifteenth floor, and revive them. Anglin promised him he would go up to the fifteenth floor next and see if Boisclair and Claire were still breathing. With that assurance, Ed Kiker allowed himself to be led onto an elevator and put in a taxi on Ellis Street to go to St. Joseph's.

When the firemen did reach the roof of the Mortgage Guarantee Building, they extended their ladders first toward the windows of the living: the Hamils and Jane Wallace. The smoke was so thick that they had not seen Ed Kiker fall.

Jane started across the ladder first. When she reached the roof, she turned to yell for the Hamils to follow her, but there was no response. Both Richard and Carlos had passed out at the window. They did not come to until after the firemen had dragged them across the ladder and taken them down to Ellis Street. Carlos was put in an ambulance to go to Grady Hospital and Richard in a cab to follow.

The Hamils were alive because they had been unable to find their own room when they were trying to escape the hall. If they had made it back to 1524, they probably would not have survived. No one else staying in the shuttered "24" rooms above the fifth floor did.

The death toll on the back side increased quickly starting at the seventh floor. There were only three victims total on the third, fourth, fifth, and sixth floors, but five of the seven people sleeping on the back side of seventh floor died. The totals climbed as the smoke rose, the fire picked up momentum, and guests were farther and farther from rescue. Eight people died on the back of the eighth floor, eleven on the ninth. In all, there were sixty victims on the back side.

Among them were six of the nine University of Alabama students who had arrived in Atlanta by train with such anticipation just five days earlier. And not only did Ed Kiker Williams' mother and sister die, but so did his Aunt Dot and three first cousins.

Trying to keep her three children calm until they could be rescued from 1530, Dot Smith had instructed Fred, Dotsy, and Mary to lie down on one of the room's two double beds. Fred, fourteen, and Dotsy, twelve, had already experienced the pain of their father's death in a car wreck, and Dot had later married Dr. Smith. Dot decided to lie down on the bed beside her children so they could draw strength from her, and she from them. Lying there together, they all died of suffocation.

Next to the Williamses' room — in 1518, the corner room that overlooked both the alley and Ellis Street — W. B. Walker survived after begging a fireman in the Mortgage Guarantee Building to rescue him next. A permanent resident of the Winecoff for more than a decade, Walker was an inventor. He created the humidifier used to spray a fine mist to hold down lint in the spinning rooms of the cotton mills that blanketed Dixie. Slightly eccentric, he would cash his royalty check at the Winecoff's front desk every month and tell the clerk to keep the change. Leaning out the window, Walker had seen day clerk Bob Tweedell in the room above him.

"Bob, we're not going to make it," Walker hollered.

Tweedell lived in 1420, and his first reaction had been to go to the roof. Having worked at the Winecoff since he was sixteen, Tweedell knew every inch of the building. He realized he wouldn't be able to see the door to the roof on the sixteenth floor because of the smoke, but he knew exactly where it was on the back hall. Once he reached the top floor, though, he was so disoriented that he headed the wrong way down the back hall. He turned a knob and found himself in 1618 with W. H. Robertson and his wife, operators of the Princeton hotels in Macon and Gainesville.

"Hang in there," Tweedell shouted down to Walker. "They're going to get us."

Then Walker had an idea. He told the fireman on the roof of the Mortgage Guarantee Building that he had fifteen people in his room. A ladder was quickly placed at his window. When he stepped down on the roof, he apologized for lying.

CHAPTER 5

꠸꠸꠸ ⚎⬩⚎ ꠸꠸꠸

D orothy Moen glanced out the window as she and her three friends from Baker Village High School finished devotionals in Room 730. She could see the huge neon Coca-Cola sign, a downtown Atlanta landmark at the confluence of Peachtree and Pryor streets. The clock on the sign read 2:30 — time for Friday to end at last.

The day had lived up to all their expectations. The girls had caught a bus from Columbus, Georgia, that morning and missed a whole day of school. Then, after checking into one of the Winecoff's largest rooms, they had spent the afternoon at the Capitol, sitting in the House chamber, participating in the opening session of the Youth Assembly, and meeting high school students from all over Georgia. They had eaten dinner at a downtown restaurant and, after the assembly's evening session, had gone with some Atlanta boys from Tech High School to get a snack at Lane's Drug Store across from the Winecoff. Rarely had the high school seniors ever felt so adult, so free.

It had been the kind of day that sixteen-year-old girls never want to end, so they had sat up talking for four hours after returning to 730. Finally deciding it was time for bed, they had their devotionals. They prayed for the family of one of their classmates, Maxine Ingram, whose mother had been killed in a house fire early Friday morning.

The four Columbus girls in 730 were among the 316 high school students who had flocked to Atlanta for the second Youth Assembly, a mock state legislature sponsored by the state YMCA to give students parliamentary experience. Elected by their classmates, they had come to offer ideas for laws that would make Georgia a better place to live, as well as give the state a glimpse of the agenda of its future leadership. The Youth Assembly delegates were all members of the Hi-Y or Tri-Hi- Y, Christian service clubs sponsored by the YMCA in high schools across America. Their goal was "Christian citizenship development."

The first Youth Assembly had been held in March, just nine months earlier. The delegates came to Atlanta, introduced bills, held committee meetings, and argued on the floor of the Capitol. Eighteen-year-olds had just been given the right to vote in Georgia, and legislators were interested in what they had to say. Governor Ellis Arnall had even spoken to them. At adjournment, a resolution was passed to hold another assembly the following year, but the YMCA later decided to schedule the assembly in December.

For this second assembly the YMCA had mailed registration forms to every Hi-Y and Tri-Hi-Y club in Georgia, asking delegates to mark their preference from a list of seven downtown hotels holding rooms for the students: the Ansley, Piedmont, Henry Grady, Atlantan, Robert Fulton, Kimball House, and Winecoff.

The girls from Baker Village High School loved their room at the Winecoff. It was large by hotel standards, with two double beds and four windows that looked out on both the Coke clock and the historic, 2,700-seat Loew's Grand, built in the 1890s. They could imagine how the streets below must have looked on a December night seven years earlier when the Grand had been the site of the premiere of *Gone with the Wind*. More than 30,000 people had gathered at Peachtree and Pryor to see Clark Gable, Vivien Leigh, and the other stars arrive and then disappear into the huge Southern mansion facade that draped the front of the Grand that night.

Dorothy woke up first, only about an hour after going to sleep. She heard someone on the sixth floor screaming for help and quickly realized the room was full of smoke, so she bolted from the bed to a window. After beating on the wooden frame around the glass, she finally got it opened. Black smoke enveloped her.

Dorothy's bedmate, Hedy Metcalf, had been sleeping closest to the door, and her first impulse had been to open it. The pressure in the hall was so great that she couldn't close it back, so the open window immediately created suction for the dense smoke seeking to escape the building. Dorothy knocked the screen out into the night air and then stuck her head out the window, looking for a fire escape.

By contrast, the Youth Assembly delegates staying three doors away in 708 — Frank Benson and Donald Davis of Quitman — didn't need a fire escape and were breathing with only minimal difficulty. The two fifteen-year-olds had been fortunate enough to get a room on the front of the seventh floor.

Earlier in the year, at his mother's insistence, Donald had read an article in *Reader's Digest* about what to do in a fire. He reacted calmly, filling the bathtub with water and then placing wet towels around the door. The two boys were among the first half-dozen people rescued by the aerial ladders. Once on the ground, they ran up Peachtree Street to the Henry Grady and sent a telegram to Quitman, deep in southwest Georgia, telling their parents they were safe.

By this time, Dorothy was perched on a window ledge in 730, trying to get away from the suffocating smoke. But she couldn't escape the smoke there either and wondered if she could make it to the room below her. As she struggled desperately for breath, the idea quickly evolved into her only real option.

With her back to the building, Dorothy squeezed the protruding bricks of the window ledge and stretched her legs toward 630. She could touch only the top of the sixth-floor windows, and her arms were beginning to ache. Her roommates were above her, gasping for air and pushing to get out the window, too, but there she hung, dangling in the darkness, helpless to pull herself back up to 730.

The night was filled with the piercing whine of sirens, as if the whole city was afire, but not a fire truck or ladder was in sight on the Winecoff's south side. Dorothy's arms had never hurt so much, and she could breathe only when a surge of wind blew away the smoke that enveloped her. She knew that if she didn't let go soon, she would lose her grip and fall. And she thought it would be better to jump than fall because she'd have a better chance of landing on her feet and not injuring her head.

"I've got to jump," she told herself.

She said a prayer, a simple prayer: "Oh, God, please don't let me die."

Then she let go. For a brief instant she felt as if she were floating like a maple leaf. Then she passed out.

She did land feet-first — with such force that the bone in her right thigh instantly snapped in two places, the lower portion popping through her skin. She had landed on the small concrete court outside Room 330, where the poker game had been raging only minutes earlier. The court abutted the flat roof of Chandler's Shoe Store. As Dorothy fell forward upon landing, her head hit the edge of the roof. Bones broke in her face. Thirteen teeth popped out.

Following right behind Dorothy, Hedy Metcalf took a flying leap out the window. She, too, landed on her feet, but when her head hit the edge of the shoe store's roof, the impact broke her neck.

Bess Dansby — only fifteen, but one of the top students in Baker's senior class — remained in one of 730's windows. She had heard one of her roommates crying but was too busy breathing to cry. The smoke was so overwhelming that she couldn't even get back to her bed, only three feet away. She realized she had to get the window closed behind her, or else suffocate from the smoke pouring through the still-open door, but she couldn't manage to close it without falling. The smoke was so heavy that her skin was turning black and so hot that she felt like she was on fire.

Finally the smoke won. She passed out and fell. Unlike Dorothy, Bess was limp when her feet hit the concrete. Still, the bone in her left thigh popped on impact. The fall snapped Bess back into consciousness, and as she lay on the terrace outside 330, she knew her classmates were

Four teenage boys from Rome and the poker players in Room 330 escaped to the roof of Chandler's Shoe Store.

there with her in that dark pit. She felt another body touching hers and could hear someone whimpering in pain.

There the three young friends lay. One was dead, another unconscious, and the third alive but unable to move. The fourth roommate, Dot Tyner, was dead in 730. And only ninety minutes earlier they had been talking as if tomorrow would never come.

If it had not been for the girls' friendship, Dorothy would have been with Betty Huguley and about to be rescued. The two had met Friday morning while checking in. Betty's father, Mason Huguley, had driven her to Atlanta after finishing his morning chores on the family's dairy farm near Sunny Side, thirty miles south. Two of Betty's classmates at Spalding High School in Griffin had also been scheduled to attend the Youth Assembly but at the last minute decided not to come. Betty had been looking so forward to the convention and spending two nights in an Atlanta hotel that she came on anyway, even if it meant staying in a room by herself. The clerk reassigned her to a smaller room on the fourth floor.

As Betty's father was leaving, Dorothy told him, "Don't worry. We'll take care of her. I'll come down and stay with her. Or she can bunk with us."

That made Huguley feel better about leaving his sixteen-year-old daughter alone in a big city. But still, as he was telling her goodbye, he gave her a warning: "As soon as you get in your room, be sure to look around for the best way out in case there's a fire."

Betty walked to the Capitol with the Columbus girls and spent part of the evening with them, but Dorothy never again mentioned their spending the night together. Betty knew Dorothy wanted to be with her friends, so she didn't bring it up. She went on to bed in 414.

Betty was so sound asleep that she could hardly understand the shouts of the man down on Ellis Street. He was drawing out the word "F-i-i-r-r-e" so long that she couldn't make out what he was saying. She could hear a roaring sound but thought she was dreaming. When she finally understood the man's cries, she realized smoke was pouring through the transom, which she had left open for fresh air. She jumped up to close it, but it wouldn't shut all the way. She dropped to the floor to get below the smoke and crawled to the wardrobe. Standing again, she calmly took her clothes off their hangers. She carefully folded the taffeta evening dress and the velvet cape that her big sister, Mona, had loaned her for Saturday night's banquet at the Ansley Hotel's Dinkler Room. With her suitcase packed, Betty turned on the light to attract attention outside. She went to the window and, using the suitcase like a battering ram, knocked out the screen.

Betty was still calm — she had never known anyone to die and didn't expect to die herself that night — when a fireman appeared at her window. Capt. Fred Brown was standing on the eight-inch cornice just below the fourth-floor windows.

"Climb out here and follow me," he said.

"Wait a minute, I've gotta get something," said Betty, turning back into her smoke-filled room.

"I'll get the baby," he said.

"There's no baby. I just need my purse."

Fred Brown helped Betty climb out onto the ledge. "Follow me," he said.

"What about my suitcase?"

"Leave it," he said.

But it was Mona's suitcase, and Betty wasn't about to return home without it. She flung it to the sidewalk and turned her back to Ellis Street. With her hair in curlers and only pajamas under her coat, she put her bespectacled face to the Winecoff's brown bricks, gritted her teeth, and began to edge her way along the ledge, toward the aerial ladder at the Winecoff's northwest corner.

At the ladder, Brown told Betty to wrap her arms around his neck. But she told him she was "kind of a tomboy" and could climb down by herself.

Once safely on Ellis Street, Betty picked up her suitcase — only the lock had broken — and headed around the corner toward the lobby, just as Dorothy, unconscious, was being carried out, her pajamas blood-soaked from her broken bones.

The lobby, untouched by fire, was packed with people oblivious to the fact that standing on the ground floor of a burning building might be dangerous. Betty sat down on a windowsill. Hoses filled the stairway, firemen were dashing in and out of the front door, and guests were standing around in pajamas and coats. There was such pandemonium inside and Betty was so mesmerized by the terror-filled scene she was watching unfold on Peachtree Street that she didn't notice Bess Dansby lying on the floor in a corner.

After extinguishing the fire on the third floor, the firemen checked each room for guests. When they looked in 330, they saw the burned-out remains of the poker game as well as the teenage girls lying on the small court outside the open windows. As he carried Bess down to the lobby, one of the firemen told her, "You're a damn fool for jumping. We would have gotten to you if you had just waited."

But the fire department rescued no one on the south side of the Winecoff that night. Fortunately, only one room on each floor faced south — the "30" room. Unfortunately, the "30" rooms — 730, 830, 930, etc. — were the largest in the hotel, the only ones with two double beds. On the lower floors, the rooms were often used as "sample rooms" by traveling salesmen wanting to show their wares to Atlanta customers. The Winecoff's management referred to the "30" rooms as their "family rooms" because they were ideal for families with children — as well as for cost-conscious Youth Assembly delegates who were paying between $1.70 and $2.00 a night to sleep at the Winecoff.

There was no reasonable escape from the "30" rooms. They could not be reached by ladders of any kind because the Winecoff's south side adjoined Chandler's Shoe Store. They were also sandwiched between what became, in effect, two chimneys. Outside their windows, a virtual flue had been created by the south side's inset design — the smoke literally hugged the bricks. On the other side of the "30" rooms, sharing an adjacent wall, was the chimney formed by the Winecoff's open stairway and elevator shafts. It sent thick, dark smoke seeping through the doorways and unbearable heat through the walls. Of the thirty-two guests staying in the nine "30" rooms above the sixth floor, only Dorothy Moen and Bess Dansby survived. Eighteen of the thirty dead were Youth Assembly participants, most of them senior girls.

The four girls in 930 were seniors at Seminole High School. It had

One of the "30" rooms, the most deadly in the hotel. Of the thirty-two guests staying in a "30" room above the sixth floor, only two survived.

taken them all day to get to Atlanta from Donalsonville, in Georgia's extreme southwest corner. The bus had broken down for two hours in Eufaula, Alabama, causing them to miss their connection in Columbus. When they finally checked in, Jeanette Riley dashed off a one-page letter to her parents on Winecoff stationary that proclaimed the hotel to be "absolutely fireproof." "I know we are going to have a good time," she wrote, then closed: "P.S. The hotel is on main street."

Jeanette jumped to her death. Senior class president Harriett Strickland, who had attended the first Youth Assembly, suffocated before fire entered the room. So did Robbie June Moye and Jerry Jenkins.

In 1130, four Gainesville High School seniors — Gwen McCoy, Frances Thompson, Ella Sue Mitchum, and Suzanne Moore — also died of smoke inhalation, then were burned beyond recognition.

Four senior girls from R.E. Lee Institute in Thomaston were found lying neatly across one of 1430's double beds, hardly a hair out of place. All had suffocated. The body of their thirty-one-year-old advisor, Mary Smith Minor, lay with them. Minor, the mother of a four-year-old daughter, had responded to student pleas and organized the Tri-Hi-Y chapter at R. E. Lee the previous year.

The delegation actually consisted of five girls — Christy Hinson, Patsy Uphold, Virginia Torbert, Earlyne Adams and Zanier Downs — and after the opening session they had come back to the room and listened to the radio built into the wall. Later they ate dinner at the Frances Virginia Tea Room, on the third floor of the Collier Building across Peachtree Street, above Lane Drug Store. It was one of Atlanta's most popular restaurants, a haven for downtown shoppers. They then went to their legislative committee meetings at Central Baptist Church near the Capitol — Christy and Patsy were on the temperance committee — and caught a trolley back to Marietta Street at 9:45. Rather than wait for a transfer, they walked the final five blocks, passing by the Winecoff to go to the Henry Grady's coffee shop for apple pie a la mode.

They were back at the hotel by eleven, where Zanier's aunt and uncle were waiting to pick her up and take her home with them for the night. She didn't want to leave, but the arrangements had already been made and she didn't want to hurt their feelings. She consoled herself with the knowledge that she still had Saturday night to spend with her friends at the Winecoff.

When the smoke awoke Minor, she opened the door just as Pauline Bault cracked the door to 1404 across the hall. "Keep your head and keep the door closed," Minor told her through the smoke. She then gathered her students together to keep them calm while awaiting a rescue that never came. A Bible was found on the bed with them. It was opened to the fourteenth chapter of John: "Let not your heart be troubled . . . In my Father's house are many mansions . . . I go to prepare a place for you."

The only Youth Assembly delegates assigned to a "30" room to survive the fire — other than Dorothy Moen and Bess Dansby — were four boys from Rome on the fourth floor. The windows of their room were directly across from, and slightly above, the roof of the shoe store. All they had to do was stand up on the window ledge and do a flat-footed broad jump of seven feet. Frank Pim knocked out the screen and was the first to jump. Roger Sumnicht and Dick Collier followed. Charles Gray glanced back as he stepped up on the windowsill. He saw the door fall in and flames shoot from behind it. He barely made it to the roof, stumbling as he landed barefooted.

Four of the poker players from 330 were already on the roof, too. They had scampered out the windows of 330, onto the concrete court, and up onto the shoe store roof. Earlier that night the boys from Rome had heard loud cursing and shouting in the room below them and wondered what was going on. Now, together, the men and boys went to the front of the roof and yelled — to anyone who might be able to hear them — that the hotel was on fire. Unable to get the firemen on Peachtree Street to send a ladder up to them, the gamblers went to the rear of the roof and climbed down a rusty, ladder-like fire escape, then shimmied the rest of the way down a drain pipe. The Rome boys followed, glancing back at Room 1030 and remembering the simple choice they had made Friday. Eight boys from Rome's Boys High School had checked into the Winecoff Friday and been given two rooms — 430 and 1030. The juniors took 430 and the seniors 1030.

Buzz Slatton, one of the seniors, was the Youth Assembly's "clerk of the house," having been elected at a YMCA summer camp at Toccoa, Georgia. He loved politics and told his friends that he intended to be governor of Georgia one day.

Youthful idealism such as his was clearly reflected in the forty-three bills submitted in November for discussion at the Youth Assembly, many of which were quite progressive for the South. Half of them dealt with health or education, and one condemned the Columbians and other "kindred hate organizations." The Columbians' attempt to gain a foothold in the state was "a slur on its good name and a reflection upon its citizenry." The neo-Nazi group's plans to bomb Atlanta's police station, newspapers, and auditorium had recently been exposed.

The teenagers were adding their voices to the national criticism that had flooded the state since the July lynching of four blacks at Moore's Ford on the Apalachee River, about an hour east of Atlanta in Walton County. The KKK had been blamed in the deaths. President Harry Truman was outraged: "When mob gangs can take four people out and shoot them in the back, and everybody in the country is acquainted with who did the shooting and nothing is done about it, that country is in a pretty bad fix from a law enforcement standpoint." On November 7, Governor Arnall had reported that fifteen to seventeen members of the lynch mob were known, but a month later there still had been no arrests.

Buzz had laboriously signed all the wallet-sized Youth Assembly membership cards given to delegates upon registration and had called the roll at Friday's opening session, which was addressed by recently elected Lt. Gov. M. E. Thompson. At the evening session Buzz supervised the assigning of delegates to the committees that were to amend, combine, or table bills. He had even had his picture in Friday afternoon's Atlanta *Journal*, posing with the other Youth Assembly leaders. Charles Keith, one of Slatton's roommates, was the younger brother of the first soldier from Rome to die in World War II. Lamar Brown, an avid stamp collector, had won the city badminton tournament during the summer. Billy Walden was president of the Hi-Y as well as of his Sunday School class. A single-wing tailback, he had just completed a football season in which he had been hailed as one of the city's finest players ever.

All any of them could do was go to the windows in 1030, try to outlast the fire and hope their friends were already out. From the windows they could see not only the shoe store roof but also — only a block away — the Piedmont Hotel, where the seven delegates from Rome's Girls High slept in safety.

Although the majority of the Youth Assembly delegates were staying in the "30" rooms, the thirteen scattered across the back of the hotel were nearly as trapped, the heat and smoke nearly as bad, the death toll nearly as high.

Three sixteen-year-olds from Albany High School — Julia Hall,

Marianne Scherle, and Ann Walker — were crowded into Room 822, thanks to a roll-away bed provided by the hotel. There they suffocated together.

On the back side of the ninth floor, seven representatives from Bainbridge High School packed themselves into three rooms. The three senior girls took 920. Ruth Powell and Maxine Willis both jumped to their death. Sue Broome, sent by the school newspaper as a press representative, remained in the room and died there. President of the glee club and frequent soloist at her church, she had recently been chosen "most popular" in the class of 1947.

Clarence Bates and Earl Carr Cragg — fourteen-year-old sophomores active in the Junior Hi-Y — were two doors down in 924. Boy Scouts and members of the school band, both suffocated.

Tri-Hi-Y advisor Mary Alice Davis, a Bainbridge teacher for twenty-two years, stayed with the two Junior Tri-Hi-Y girls in the delegation — Mary Lou Murphy and Patsy Griffin — in 926. They were attending the assembly simply as observers, preparing for the opportunity to be delegates as seniors. Mary Lou was president of the sophomore class and a natural leader, and Patsy had an innate interest in government. Her father, Bainbridge newspaper publisher Marvin Griffin, was state adjutant general, essentially the commander of the Georgia National Guard. The two girls died, as did their advisor, who was also youth director at Bainbridge's First Methodist Church.

Two floors directly above Miss Davis and the girls, Rudolph Ogilvie and Bobby Sollenberger were spending their first night ever in a hotel. They had caught the bus from Barnesville Friday after getting out of class at Milner High School. They took a taxi to their assigned hotel, the Henry Grady, but the desk clerk told them he had no room for them and they would have to go to the Winecoff. They were assigned 1126. They met the girls from Gainesville, who were staying in 1130, and also talked to the Thomaston girls, whom Rudolph already knew.

Bobby knew to wet towels when he smelled smoke. He threw one to Rudolph.

"Here," he said, "breathe through this."

But as they looked out the window, the heat in the room became unbearable. They were blinded by the smoke and darkness, and sweat was running in their eyes.

"I'm going to jump," said Bobby. "There's a net down there."

Rudolph looked down into the alley. "That's not a net, Bobby," he said. "And I don't think jumping is a good idea."

"Well, I think we're going to burn to death up here," Bobby replied.

"Well, we might, but I'll burn before I'll jump. You'll die for sure that way. At least we have a chance this way."

Rudolph began to feel his way toward the door fifteen feet away, stumbling over the furniture in the heavy smoke. When he did locate a door, it opened into the bathroom, not the hall. Running his hands along the wall, he finally found the door to the hallway and opened it. He could see small flames beginning to burn the hall carpet.

"Come on, Bobby," he called behind him. "I can see the stairs. We can go to the roof."

"Don't leave me, Rudolph," Bobby begged.

"I'm not going to leave you, Bobby."

Rudolph heard Bobby fall and turned back into the room to find him. He fell over furniture he had already knocked over, got up and fell over more furniture, searching futilely for his friend. With the sweat and smoke burning his eyes, it was impossible to see anything.

"Bobby," Rudolph called out repeatedly, "where are you?"

After searching the room as long as he could stand the smoke, Rudolph headed back toward the door. He knew he, too, would soon pass out if he didn't get out of the room. Rudolph glanced toward the window and, through the smoke, saw flames leaping out of the window below. He turned back toward the door. He would have to make it to the roof by himself.

He headed into the hall, bare-chested, wearing only the undershorts he had been sleeping in. He quickly realized that the stairs that had looked so close went down, not up, so he headed on down the hall, stepping around the flames on the carpet. He could see well enough to find the stairs going up to the twelfth floor between the elevators. Inhaling the smoke and poisonous gases pouring up the stairway,

Rudolph fell to his knees and began crawling up the steps.

He made it to the twelfth floor, but because the Winecoff's stairway was T-shaped, he realized he would again have to go down the hall and around the elevators to get to the stairs to the fourteenth floor. He didn't think he could make it up another flight. He saw a half-open door along the back hall, pushed on it, and stumbled across the threshold.

The room was pitch dark. Rudolph tripped over something on the floor — the body of a woman. He fell forward and landed face down, on his nose. A voice spoke out of the darkness.

"Did you close the door?"

"No," gasped Rudolph, unable to get up.

"Oh, hell," said the voice.

Rudolph was unconscious but still breathing when a fireman carried him across a ladder to the Mortgage Guarantee Building.

Rutledge Minnix didn't have the chance to get to a ladder. He was fifteen years old, a junior at Columbus High School, and by himself at the Winecoff. Before going to bed, he had kissed his mother good night on Peachtree Street. She had come up from Columbus with him, despite his pleading that she let him come with the seniors from school. He was her only child, she told him, and she was worried that something would happen to him.

"You're trying to make a baby out of me," he told her.

His father told him, "Your mama is just an old hen with only one chick, and she's afraid something will happen to that one chick. She's not going to embarrass you."

Finally Rutledge relented and said she could come.

The Minnixes owned a seven-acre nursery — known throughout the South for its camellias — that surrounded their antebellum home in the heart of Columbus. Rutledge was a brilliant student, possibly headed for an Ivy League education. He was interested in everything, from debate club to student council. When he was eleven, he had taken a map of Columbus and laid out plans for both attacking and defending it.

When he was twelve, his life was touched by catastrophe when he accidentally shot and killed the daughter of his parents' best friends.

They had been playing cops and robbers at Rutledge's house when he picked up the pistol that stayed in a bedside drawer. She ran by him, and he put the gun to her back and fired. In the three years since, he had never mentioned the shooting. It was as if his mind had erased something too terrible to remember.

Because Rutledge's room at the Winecoff — 1116 — was so small, Ruth Minnix took a room at the Henry Grady, one block up Peachtree. The four senior boys from Columbus High were at the Grady, too. They had requested the Winecoff but mailed their forms in too late. Four Columbus High girls were staying at the Ansley.

While Rutledge attended the Youth Assembly session Friday evening, his mother went to a movie. Afterward, he came by to see her. She walked with him back to the Winecoff.

"Please get your suitcase and come stay with me," she told him.

"Now, Mother," he replied, "you promised you wouldn't take charge. For once, I want to do something by myself."

They paused to look through the window of Henri Monet's perfume and jewelry shop on the first floor of the Winecoff — at a clock they planned to buy Rutledge's father for Christmas. Rutledge gave his mother a kiss and then went up to his room to work on the speech he was to give Saturday.

When his mother was awakened by the sirens, she called the Grady's front desk. "You can go on back to sleep," the clerk told her. "That fire's down the street at the Winecoff."

She ran all the way there, frantically questioning firemen, policemen, bystanders — anyone she could — asking if they had seen her son. Then she began pleading hysterically with the firemen that they let her go up into the hotel and try to save him. Overcome by smoke, Rutledge was already dead.

In all, thirty of the forty Youth Assembly delegates staying at the Winecoff were killed by the fire. The bodies of twenty-six of them were burned, most beyond recognition. Of the ten survivors, five had been fortunate enough to be given fourth-floor rooms. Two more had been assigned a seventh-floor room on the Peachtree side, from where they were quickly rescued. The other three — Dorothy Moen, Bess Dansby, and Rudolph Ogilvie — had escaped miraculously.

Fireman Roy Buford carried Dorothy out of the hotel. Buford had been standing in Room 330 — scene of the poker game — when he heard two thuds in rapid succession outside the windows. He looked out and saw Dorothy, quickly picked her up, and carried her down Peachtree Street toward an ambulance. Dorothy came to enough to tell him her name and address. She asked him if she would live. He said she would.

"I love you," she murmured through missing teeth.

CHAPTER 6

U pon checking in late Friday afternoon, Reid and Cary Horne boarded one of the Winecoff's two elevators to go to their room. As they
passed one floor after another without stopping, Cary remarked,
"Good gracious, we're going up so high."

"Yes ma'am," said the bellboy holding their bags, "to the sixteenth
floor, the top of the building."

"Well, I hope we're near a fire escape," she replied.

"Oh, we don't have fire escapes, ma'am. This hotel's fireproof."

The Hornes — like Ed Kiker Williams — were from Cordele. In fact,
they were distant cousins of the Williams and Smith families and rented a
house from Ed Kiker's grandmother. But the Hornes had no idea that the
Williamses and Smiths were staying at the Winecoff only one floor below
them.

The Hornes had come to Atlanta for surgery, just as countless
Georgians did every week. Many people considered the hospitals and other
medical services in the state capital superior to those in their smaller home-
towns. Reid, a produce broker, was to have a tonsillectomy early Saturday

morning. After he and Cary went to dinner, they played cards on their bed in 1606 before going to sleep around nine o'clock. They had to be at the hospital early.

They awoke to screaming that sounded like it was in the hall. Reid cracked the door to take a look. The pressure of the smoke at the top of the building was so great that Cary had to help him close it. By then a thundercloud of black smoke had billowed into the room. They began to dress and fill the tub with water, but everything they did seemed in slow motion. They repeatedly ran to the windows for fresh air, but even there breathing was difficult. The wind was blowing smoke from all across the front of the building into their faces. They knew they would not be able to wait the fire out in their room. The pressure in the hall was too intense, and the air at the window too smoke-filled. They were trapped.

Directly below the Hornes, in 1506, J. W. Henry was looking out his windows also. He called up to them.

"Do you know any way we can get out?"

"I'm sure there's some way to get to the roof, but I don't know how," Reid shouted back. "The hall's full of smoke. We wouldn't have a chance."

Then Reid had an idea. He had noticed that, with the wind blowing from the northeast, there was little smoke on the north corner of the hotel's front side. He had also noticed that less than two feet below the fifteenth-floor windows an eight-inch cornice ran across the front of the building, setting off the stucco above. It was just like the decorative cornice below the fourth floor windows that topped the stucco covering the bottom three floors.

Reid thought he and Cary could escape the smoke by walking the cornice to the north corner, toward Davison's. They were young — Reid was thirty-two and Cary thirty — and in good shape. He believed they could do it.

"Will you help us down into your room?" Reid yelled to Henry. "We want to walk the ledge over to that corner room."

Cary went out the window first, and Reid interlocked their wrists to let her down slowly. Henry, who had pulled his window down from the top, grabbed her by the waist and pulled her into 1506.

Then it was Reid's turn. Clutching the window ledge, he lowered himself until his legs were in front of Henry's window. Henry told Reid to let go as soon as he, Henry, grabbed his legs, that he would pull him into the room. But Reid was a big man — six-feet-three and 220 pounds — and he knew that letting go wasn't a good idea. What if Henry couldn't hold him by the legs well enough to pull him in?

So, with one hand still squeezing the window ledge in 1606, Horne reached down with his other arm and grabbed the top of the window frame in 1506. With Henry holding his legs, he pulled himself into 1506.

The smoke in 1506 was just as bad as it had been in their room, so the Hornes climbed right back out the window — ready to make their walk along the cornice. The woman in the corner room had yelled to them that the smoke wasn't too bad in her room, that they should come on.

"Don't look down," Reid told Cary. "Just concentrate on where you're headed. Keep your eyes on the bricks and dig in with your toes."

Cary was wearing a bulky fur coat she had borrowed from Reid's older sister. Reid was wearing slacks and a T-shirt. Cary was barefooted; Reid was wearing socks but no shoes.

It was a spellbinding sight for the crowd on Peachtree Street to watch. The Hornes looked like two small, spider-like shadows creeping along the face of the building, their arms spread to clutch the stucco above the cornice. Step by step they slid, small step by small step, easing along ever so gently toward the hotel's northwest corner.

The Hornes needed to reach only the next set of windows, as 1508 was part of a three-room suite that stretched all the way to the corner. But they didn't know that; the woman on the corner hadn't told them. So they kept heading for 1512. And made it.

From Room 1504, Nell Sims watched the Hornes' daring walk across the cornice. The heavy smoke was blowing right into her face, and she wondered if she could make it, too. She was a thirty-three-year-old secretary for Aldora Mills, a subsidiary of General Tire and one of the countless cotton mills that were the economic anchor of small towns across the South. Just a week earlier, she had received the charter of the recently formed Business and Professional Women's Club in Barnesville, Georgia. The club's state president had presented it to her as a reward for having helped organize the chapter and being elected its first president.

Sims had come to Atlanta at the invitation of Reeves Marquis, her childhood friend. Marquis was from Bartow, Florida, but her grandmother lived in Barnesville, and she had visited often while growing up.

Marquis wanted to come to Atlanta to see her boyfriend Friday night. She also wanted Sims to help her shop for her trousseau. Her mother told Marquis she could go only if Sims went with her. Sims' mother didn't want her to go, but Sims had insisted.

The two women picked Marquis's boyfriend up at the airport, and he and Marquis later went out. When Sims awoke to the smoke in 1504, Marquis was still not back. Sims assumed she was still with her boyfriend, who had a room on one of the Winecoff's lower floors.

One of Sims' first impulses was to try a door that opened into 1502 — the two rooms could be rented as a suite — but she couldn't open it. Neither could the two people in 1502 — Lena and Anne Smith. Each time one of the women would unlock the door on one side, the person in the other room would relock it.

The Smiths, from Valdosta, Georgia, had come to Atlanta to shop for Christmas. Anne's cousin, Mitchell Smith, had driven them up Friday. Her father, Frank, had stayed home to run his electrical supply business.

Anne, fourteen, was Lena and Frank's only child. Their first two children had died as infants. Then when Anne was three, they discovered she was more than eighty percent deaf. She went to a boarding school for the deaf in Savannah and learned to read lips well enough to return home to a small Catholic school that offered the individual attention she needed. When she entered Valdosta High School, teachers devised ways for her to "hear" the lessons. She had gone to camp at Pine Mountain in the summer of '46 and was a member of both a high school sorority and the Tri-Hi-Y.

Eight other Tri-Hi-Y members from Valdosta were supposed to be staying at the Winecoff that night, too, but there had been a mix-up in reservations. Their chaperon knew the manager of the Ansley and pleaded for help. He gave them one room and then squeezed seven cots into an adjoining salesman's sample room.

Anne and Lena spent the evening watching *Song of the South* and were asleep well before midnight. Awakened later by smoke entering 1502, they tried to close the transom but couldn't shut it completely. As the room quickly filled with smoke, they covered their faces with wet towels and went to the window. Like Nell Sims, they soon realized that if they didn't follow the Hornes across the ledge and away from the smoke, they might die.

Lena, however, was a large woman and could not squeeze over the low iron grating at the bottom of the window, which was designed to discourage fifteenth-floor guests from crawling out onto the cornice. She collapsed, exhausted, on the radiator. Unable to revive her mother, Anne faced a terrible decision: leave her mother, possibly to die, or try to get help. She climbed out the window.

So did Nell Sims, who had pulled her winter coat over her nightclothes and tied her hair up in a scarf. She was on the ledge only briefly before losing her balance. She landed on the hood of a firetruck and broke her neck. Somehow, Anne maintained her composure well enough to make it to the corner suite.

The suite was the residence of the family of Art Geele, Jr., one of the hotel's operators. Geele and his father, A. F. Geele, Sr., had taken a lease on the Winecoff two years earlier. It had been their big break. Having operated the four-story Gilcher Hotel in little Danville, Kentucky, they had accumulated enough money to purchase the Winecoff's lease in 1945. The lease had become available when its longtime holder, the Meyer chain, relinquished it, believing the hotel not to be in a good post-war location.

The Geeles had purchased the lease from Atlantan Annie Lee Irwin. Mrs. Irwin had bought the hotel in 1943 from King Investment and Security Company for $600,000, assuming loans totaling $480,000. Mrs. Irwin was a Fairburn farm girl who worked in her father's grocery store for a decade before marrying Walter Irwin. An eccentric man, Irwin owned a small airport at Fairburn — southwest of Atlanta — as well as the Hangar Hotel near the fast-growing Atlanta airport. He was known by pilots as "The Hip Pocket Banker." Irwin bought the Winecoff as an investment and put it in his wife's name. She was the quintessential absentee owner. She had visited the Winecoff only twice and never ventured above the mezzanine.

The Geeles had signed a ten-year lease as the Winecoff Hotel

Company. Art Sr. was the president and Art Jr. the hotel manager. A third partner — Chicago policeman Robert O'Connell — supervised the hotel's work force, served as house detective, and did whatever else the Geeles asked.

One of first changes Geeles and O'Connell made was to fire L. O. "Lasso" Moseley, the Winecoff's manager for ten years. An Atlanta alderman, Moseley had offered to stay on at the hotel to help with the transition — without pay, since he had received a large severance from the Meyer chain. But the Geeles and O'Connell refused his help. The move was seen as a curious one in the Atlanta hotel community, as the new lessees were generally considered unprepared for running a hotel the size of the Winecoff. Moseley, meanwhile, moved on to manage the more prestigious Henry Grady.

Art Sr., a distinguished-looking man with the chain of a pocket watch always dangling from his suit, appeared very much the hotel tycoon sitting in the Winecoff's lobby. As a young man, he had been mayor of Sheboygan, Wisconsin. He was in the hardware business there before moving to Chicago, where Art Jr. was born.

Art Jr., the general manager, had been in the hotel business for a decade. He was a graduate of Cornell University's esteemed school of hotel management and also a trained accountant. He ran the Winecoff like he did the Gilcher Hotel back in Danville — by trying to please everyone. The Geeles had been looking for an opportunity like the Winecoff, and they were determined to make it go. They were putting their earnings back into the hotel.

But on the night of the fire, neither Art Jr. nor Art Sr. was in the hotel. They had gone deer hunting in southeast Georgia, at Dover Hall Plantation near Brunswick. They were guests of Allen Chappell, a Winecoff resident and member of the state public service commission. So Art Jr.'s wife, Gladys, was glad to have the company of the Hornes and the others in the family's fifteenth-floor suite. Gladys had fallen in love with Atlanta and was busy making friends and trying to get involved in civic activities. Five feet tall and strikingly attractive, she had been a beautician in Danville when she met Art Jr.

Since being awakened by the fire, Gladys had been shouting up to sixty-year-old Maude Whiteman, who was directly above her in the six-teenth-floor suite where Gladys's in-laws lived. Art Sr. had asked Whiteman, the cashier at the cigar counter in the lobby, to stay overnight with his wife, Esther, who was confined to a wheelchair because of arthri-tis. Whiteman had hollered out the window for Gladys and her two chil-dren — four-year-old Esther and fifteen-year-old Johnny — to come upstairs and help her take their grandmother up to the roof. But Gladys told her that there was no need to go to all that trouble.

"The fire department's already here, and the fire will never make it up this high," she yelled. "It's down on the third and fourth floors, and the hotel's fireproof anyway."

But shortly after the Hornes and the others climbed into the safety of 1512, flames began erupting on the north end of the hotel's Peachtree side. Another flashover had occurred, and orange fire was popping from the windows several stories below.

By this time, Robert O'Connell had stepped through the window of 1512. Having seen the others walk the ledge from 1602, he used a sheet to drop to the Smiths' window ledge, then stepped down on the cornice and walked the length of the building.

A strong, stout man, O'Connell was a curious addition to the Geeles' management team. He was on a leave of absence from the Chicago police force while his wife and daughter were still living in Chicago. He had no hotel experience, other than walking a beat around Chicago's Drake Hotel. Yet he had been given supervision of the Winecoff's 120 employees and of hotel maintenance. He was a partner not only in the Winecoff but also in the Geeles' other hotels — the Gilcher in Danville and the Central Hotel in Macon, Georgia. O'Connell was an unpolished man, and the suits he wore could not hide his rough, crude nature. The Winecoff's employees didn't like him. Nor did the patrons of the hotel bar, the Plantation Room, from which he would literally drag out drunks, haul them down the stairs and through the lobby, and then physically throw them out onto Peachtree Street. He would not hesitate to cancel a guest's reservation to take care of a friend who wanted a room.

To the gathering crowd in 1512, O'Connell seemed drunk. He usually roamed the hotel until dawn, but on this night he had retired early. He claimed to be sick with the flu, but the Hornes and J. W. Henry doubted it. Certainly, O'Connell was in no condition to participate in the decision-making to take place in 1512.

There were nine people in the room — O'Connell, the Hornes, Henry, Gladys Geele and her two children, Anne Smith, and Gladys Mitchell, another fifteenth-floor guest. The room was becoming increasingly smoke-filled; the flames below were moving steadily toward them, one floor after another. It was obvious that the fire department was having difficulty stemming the flames' advance, and they wondered if the fire would keep moving upward and trap them.

The Hornes noticed that no smoke was coming from the rooms above them, and Maude Whiteman told them there was very little smoke in the suite. Although she had initially wanted to take Mrs. Geele to the roof and the stairs were only a few feet down the hall, Whiteman knew she couldn't pull the wheelchair up the stairs without Gladys's help. Figuring she would be forced to stay there, she had sealed the door and transom tightly.

Whiteman tied several knots in a sheet rope to make it easier to climb, then tied it to the mullion between the windows and dropped it toward 1512. Henry said he would try it first. When he finally pulled himself into 1612, he was so choked from the smoke he had breathed during his climb that he collapsed exhausted in a chair.

Gladys Geele went next. She managed to climb up to the sixteenth-floor window but couldn't pull herself into the room. Whiteman reached for her, but Gladys was like a drowning person, flailing her arms, unable to help herself. She nearly pulled Whiteman out the window. Whiteman wrapped her left arm around the mullion between the windows and extended her right arm. She pulled Gladys into the room but wrenched her back in the process.

Down on the fifteenth floor, a sort of basket was constructed from the sheet rope for Gladys's children. Little Esther started crying when told she couldn't take her bride doll with her. Reid stepped out onto the window ledge to help push Esther and her older brother up to the next floor. Anne Smith, Gladys Mitchell, O'Connell, and the Hornes then followed. As the crowd below on Peachtree Street watched in amazement, nine people safely ascended from the fifteenth to the sixteenth floor, their bodies scaling the stucco upper face of the Winecoff.

The firemen couldn't catch up to the flames. They were running up and down the stairs, which were buried beneath more than a dozen hoses. They stood in water above the knee, operating their heavy hoses until relieved by other firemen. On their way out they would help a guest or carry a body. As soon as they caught their breath outside, they reentered the brick inferno.

The lines kept advancing, slowing only when the slack ran out and a hose had to be reconnected to a standpipe outlet or when a floor was taken and the firemen had to search for victims. They went down the smoky corridors, through the rubble of fallen plaster, never sure when they would step on a body, always wondering what lay behind the next door. The smoke disoriented them. They could neither breathe nor see well and were often unable to find their way. A buddy would lead them out, and the crowd on Peachtree Street would see them stumble as they emerged from the building. Only a few wore canister masks. They could breathe through them for only a few minutes, and pulling the cannister was difficult. The 2.5-inch hoses, with a double cotton jacket over a rubber lining and brass couplings, were heavy enough. Filled with water, fifty feet weighed 112 pounds.

The eleven people now in Suite 1608-10-12 had little difficulty breathing, but they were growing increasingly worried. As they looked out the windows, they could see that the flames were still climbing floor by floor. Before passing out, O'Connell had told them there was no way they could get through the hall to the stairs to the roof. He had tried earlier, and the smoke was too thick.

Johnny Hall — Gladys Geele's son from a previous marriage and a freshman at Atlanta's Boys High School — was getting wet towels for everyone and trying to keep his family calm. Reid Horne was emerging as the group's leader. Anne Smith asked him several times about Lena, her mother. Although Anne had difficulty pronouncing words correctly — because she couldn't hear them — Reid understood her well enough to know she wanted help for her mother.

"Is she a big woman?" Reid asked.

She read his lips and nodded.

"Well, she's better off where she is," he said.

After about twenty minutes in the room, watching the rising flames, the group became convinced that the fire department was failing to contain the fire. A sheet rope beckoned to them from 1112, four floors below. Catherine McLaughlin had used it to drop to a sheet rope at the ninth floor and then to a ladder at the seventh floor.

"If we could make it to that rope down there, then we could probably make it to the ladder," Reid said.

The Hornes, Henry, and Gladys Mitchell began to wet sheets, knowing that would make them stronger. Having walked an eight-inch ledge and climbed a sheet rope, they would rather again test their physical strength than risk dying where they were. They tied their sheet rope to the radiator below the window.

The first person out the window was Gladys Mitchell, a J. C. Penney employee in Maryville, Tennessee. She didn't hesitate when Reid offered her the rope. Wearing a black fur coat over her nightgown, she kicked off her shoes and crawled onto the window ledge. Reid planned to let the women go first, and then anybody who wanted could follow. He certainly wasn't going to let too many people get on at one time; in the past half-hour they had all seen overloaded ropes snap.

"I've got to make it," Gladys Mitchell kept saying to herself, over and over. All her life she had been able to maintain calm in the midst of danger. She had grown up in the east Tennessee countryside, the only child of a building contractor who had pushed her to be strong and independent.

Gladys descended a floor and then Cary took the rope.

"Stop on the window ledge at every floor so you can get a good grip again," Reid told her. "And you two alternate. Only one person on the rope at a time."

Cary made it to the fifteenth floor fine. Then while Gladys paused on a window ledge below, Cary lowered herself to the fourteenth floor. There, standing on the window ledge of 1412, she could hear the voices from 1612. The group was trying to decide who would go next. She heard Reid offering them the sheet rope, and she looked up to see Anne Smith climbing out the window.

People on Peachtree Street started hollering, "No! No! No!" They too

Reid and Cary Horne and nine other people were trapped in 1612, the top room on the right corner of the hotel. They tried to make it to the long sheet rope hanging from 1112.

had seen overloaded sheet ropes snap.

Anne was not an athletic person, and her heart was heavy with worry for her mother. She took the sheet from Reid once she was on the window ledge.

Cary was still looking up. It seemed as though Anne didn't have a good hold on the rope; then Cary felt Anne's body brush her hand. Gladys Mitchell was dangling on the sheet rope when Anne's body hit her on the head. Both hurtled toward the pavement. Reid could hear Anne's body hit the street. Gladys Mitchell didn't make it that far.

Gladys's left arm struck the cable supporting the vertical Winecoff sign on the corner of the hotel. The cable snapped on impact, instantly wrapping around her arm like a vise. The cable held firm to the building, though, and Gladys twirled like a leaf more than twenty feet above the sidewalk. By preventing her arm from being ripped off, her black fur coat had most likely saved her life.

Cary, still perched at the window on the fourteenth floor, couldn't see what had happened to either Gladys or Anne.

Reid hollered down to her: "We're going to pull you back up."

Reid climbed out the window to come down after his wife; he realized they had misjudged how far away the ladder was. Also, the flashover had erupted as soon as Gladys started down the sheet rope. Flames bursting out of the Winecoff's tenth-floor window burned the lowest portion of the sheet in two while Reid and Cary stood at the fourteenth floor. Now the Hornes wanted to get as far away from the fire as possible.

The flames were raging so on the eleventh and twelfth floors that the plaster was falling off the walls in chunks. Assistant fire chief Fred Bowen — who had called for help within two minutes of getting to the Winecoff — was being dragged out of the building for the second time. He had been overcome with smoke and hauled out earlier, then came to and reentered the building.

The firemen had placed deluge guns — essentially water cannons — on Peachtree to soak the front of the building. Other firemen were climbing ladders with hoses in hand and shooting water into windows above them. But all the gallons of water on the outside seemed a terrible mismatch against the raging inferno inside the building. The orange glow

Gladys Mitchell's fall was halted by the cable supporting the vertical Winecoff sign on the corner of the building. It snapped, wrapped around her arm, and suspended her above the sidewalk.

looked like the inside of a furnace.

The underground water pipes buried during the nineteenth century were carrying thousands of gallons per minute to one of the highest points in downtown Atlanta. Chief Charles Styron was not sure how long the pressure could be tolerated. The city's entire infrastructure was under strain. The telephone and telegraph systems were operating under their heaviest loads ever.

Out of the dark, though, help was crowding onto Peachtree Street. Boy Scouts, Gray Ladies, and the Red Cross were setting up stands to serve coffee to the firemen and other rescue workers and were making sandwiches at Lane's, the drugstore on the corner across the street. The store had been opened by Georgia Tech student Arnold Hardy, who had hurried to the Winecoff and to shoot some pictures. He noticed a fireman and a policeman reading the emergency card on the door of the drugstore and talking about calling the manager to come open it for medical supplies.

Boy Scouts rushed to Peachtree Street to set up coffee stands for firemen and other rescue workers.

Hardy told them to break the door open themselves, then kicked it in when they refused. The policeman immediately arrested him for disorderly conduct. But now Winecoff guests with minor injuries were being treated on cots in Lane's, crying for news of loved ones.

— • —— ⚔ ◆ ☰ — • —

The crowd on the street saw the Hornes scale the building on their way back to safety. Once she was back in the sixteenth-floor suite, Cary lay on the floor exhausted. The mood in the room quickly turned dismal. The remaining nine knew they were stranded now, as if on a vertical island, able to see hundreds of people but be reached by none of them. Death was all around them. They believed they were the only ones still alive on the Winecoff's top floor.

The only other guests on the Peachtree side of the sixteenth floor — the Parkers of Montgomery, Alabama — were in 1602, on the other end of the front hall. Margaret Parker and her twelve-year-old daughter, also named Margaret, had ridden the train to Atlanta to go Christmas shopping and stay at the Winecoff, where the Parkers had honeymooned. Her husband, Charles, had stayed home to run the family furniture store.

When the fire broke out, the Parkers had gone into the hall, where O'Connell heard them and let them into his room. But they had breathed so much smoke that they both passed out. Margaret fell back on the bed and died. Her mother collapsed face-down on the floor and was still breathing, as most of the room's oxygen was at floor level. O'Connell, meanwhile, had fled the room through the window.

In an adjacent room, 1630, the Dickerson family lay dead, huddled together in the tiny bathroom.

Will Dickerson, a graduate of Washington & Lee law school, had survived the Battle of the Bulge just two years earlier. During that winter of 1944-45, he and several other soldiers had been cut off from their unit and forced to hide in an unheated cellar. He nearly lost his feet to frostbite.

After his release from the Army in 1946, Will returned to his job as director of unemployment compensation at the state labor department. He and Mary decided to build a house in Jonesboro, south of the city. He

111

lived in Atlanta while it was being built, and Mary and the two children stayed in Douglas.

Will went home for Thanksgiving, and on Sunday he brought Mary and the kids back to Atlanta with him. Their house was nearly completed, and they were scheduled to move in by Christmas. They checked into one of the Winecoff's "family rooms."

The fire forced them into the bathroom for water. Dickerson tried to calm his family as the rising heat poured through the wall between the bathroom and the stairwell. A vent — sandwiched between the bathroom and the hallway linen closet — sucked poisonous gases into the small room. Dickerson died with his head in the shower, Mary with an arm around each child.

Death reigned over the back hall of the sixteenth floor.

Louise Brown and Mary Stinespring, the two good friends staying in 1628, adjoining the Dickersons' room, had died in the hall. Both widows, they lived across the street from each other in Cornelia, in northeast Georgia.

Brown had a doctor's appointment in Atlanta and had asked Stinespring to go with her. They agreed to turn the trip into a small Christmas shopping excursion as well and were planning to stay at the Henry Grady. Cecil Cannon, Stinespring's first cousin, was one of the Grady's operators, and he could usually figure a way to get her a room. This time, though, he couldn't, but he did offer to call around to find other accommodations.

"The Biltmore and Winecoff both have rooms," he reported. "Which do you prefer?"

The Biltmore — on West Peachtree — was several blocks north of the downtown shopping district, so Mary requested the Winecoff.

The two women were given Room 1628, on the back corner of the top floor. When the fire started, they left the room in search of a fire escape or a way to the roof. They didn't make it out of the hall.

Next door in 1626, Bill Berry — a twenty-five-year-old ex-Marine who had fought his way through Guadalcanal and the South Pacific — was dead. A bartender at Jack Sheriff's Theatre Restaurant next door, he was living at the Winecoff, a step up from the cheaper hotel where he had been staying. He had come to Atlanta as soon as he returned from the war; there was no money to be made back home in Cedartown, Georgia. His parents had given up cotton farming before the war and were running a little country grocery store. His big brother had gone to work at the mill. Berry liked the excitement of the city. When he wasn't tending bar, he bowled — either in tournaments around the city or one-on-one with a wager on the side.

Next door in 1624, Jimmy McDonald also lay dead from suffocation. Just fifteen months earlier, he was on a destroyer in Japan when the treaty was signed to end the war. A CPA for the Federal Public Housing Authority, he was stationed in Atlanta but spent most of his time flying to sites in his five-state southern region. He got home to Memphis — and his wife and seven-year-old daughter — about once every three weeks.

McDonald, thirty-nine, went home for Thanksgiving and returned to Atlanta on Sunday. He usually stayed at the Piedmont Hotel, but it was full. When he asked where he should go, the desk clerk suggested the Winecoff. He wrote to his wife, Dorothy, Friday night, just as he did every day. He told her he had spent the evening watching bellhops put drunks on the elevators in the lobby. "It's a wonder they didn't set the place on fire," he wrote.

In 1622, Langdon Thrash was unconscious, his head lying on the window ledge with a towel around it. Before passing out, he pulled the window down on his shoulders so he couldn't draw his head back into his smoke-filled room.

In 1620, Mary Helen Rasmussen was also clinging to life at the window, but her husband, Carl, and twenty-year-old son, Richard, had already suffocated. The Rasmussens had lived at the Winecoff for a month, waiting to move into the house they had bought in Atlanta. They were supposed to take possession the following week.

Carl worked at the Veterans Administration Hospital in Atlanta. A graduate of Northwestern University, he was a small-town doctor in Illinois and Iowa before moving to Des Moines during the war. He joined the growing VA staff in May and was transferred to Atlanta in the fall.

Mary Helen was eventually rescued by Grady Hospital intern Bithel Wall, who took her across a ladder to the roof of the Mortgage Guarantee

Fireman R.L. Ellington rescued guests by crawling across ladders from the roof of the Mortgage Guarantee Building.

Building. He had been summoned to the roof by Jimmy Cahill, who, after rescuing his mother from 620 and sending her to the hospital, had turned his attention to the Winecoff's upper floors. That seemed to be the only place where anyone was still alive. He had gone down to the alley to get two ladders that had been used to rescue guests on the lower floors. There, on the sidewalk on Ellis Street, he had seen Wall and another intern, Joe Hooper, and asked them to help.

Both interns were fresh out of medical school. Wall had just finished at the University of Georgia, where he was president of the student body, and Hooper, a native of Wilmington, North Carolina, had graduated from Harvard. Both had been in the Winecoff earlier that night, at the intern party in Sonny Baird's room on the eighth floor. Not long after getting back to their quarters at Grady, they were alerted to the fire and took a Grady ambulance to the Winecoff as soon as they could. They wanted to see if they could help Baird or his wife, Elsie, or Wall's girlfriend, Helen Rutledge, who had spent the night with the Bairds. After seeing Baird and the two nurses make it down to Peachtree Street, Wall and Hooper had gone around to the back of the building.

After Wall helped Mary Helen Rasmussen down the ladder — the roof was across from the Winecoff's fifteenth floor — he and Hooper retrieved three more bodies. Cahill was trying to recover a body as well, when a young fireman named Red Pittman came to his aid. Pittman had been an Army fire marshal on Oahu during the war.

On orders from Station 4 captain Big Boy Anglin, Pittman also rescued Annie May Constangy, who was screaming for help from 1528. Her husband, Morris Constangy, had just died in her arms, and she knew her son and his wife were probably trapped on the twelfth floor. When Anglin stretched a ladder from the Mortgage Guarantee roof toward her room, it didn't quite reach. So he sat down on the edge of the roof, propped the ladder at a forty-five-degree angle against his expansive stomach, and called for Pittman to crawl across the ladder.

The Constangys had been permanent residents of the Winecoff since moving there in September. With their children grown, Annie May was ready to give up housekeeping and be free to travel with Morris, a sales representative for a dress manufacturer. He was the son of Russian immigrants who had come to Atlanta in the nineteenth century.

Their son, Joe, had just bought a dime store — Sims Five and Ten Cent Store — on Boulevard, across from Georgia Baptist Hospital. He and his wife, Freida, who was three months pregnant with their first child, had rented an apartment near there. Although the water had not been turned on, a new mattress and box springs had been delivered, and Joe thought they could go ahead and start sleeping there.

"No, come stay with us at the Winecoff," Annie May had told him that week. "We'll pay for it."

Joe and Freida checked into 1222 Thursday.

Annie May didn't want to go across the ladder with Pittman. She was near hysteria and convinced she couldn't make it. Pittman grabbed her under the shoulders and pulled her across — head first.

Down below, suffocation victims from the Winecoff's back side were lying on the sidewalk of Ellis Street. Firemen turned their hoses on them in a last hope of reviving them. Nearby, undertakers jockeyed and argued for the opportunity to get the next body. They stood in line with their hands in their pockets, their wheeled stretchers lined up beside them.

On Peachtree Street, the Winecoff's awning was now in tatters, rent by falling bodies. A beam from the fire department's floodlight truck eerily framed the smoking building. The light made the mist coming off the hoses glisten and reduced human figures to weirdly distorted silhouettes.

* * *

The stream of light was focused on the northeast corner of the top floor, where the Hornes and the seven others were still trapped. Each time one of them would appear at the windows of 1612, a fireman with a megaphone would shout them back. The crowd on Peachtree Street had seen enough people jump already. Not one of the nine people in the suite was

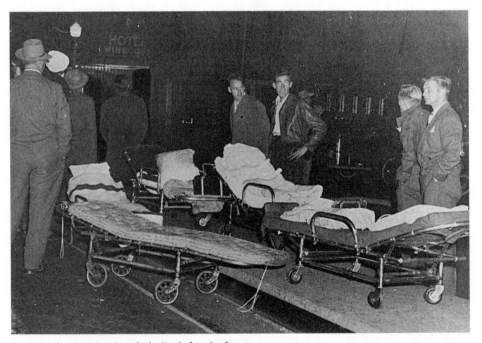

Undertakers lined up for bodies before the fire was even out.

talking about jumping, but several of them were thinking about it. The Hornes had agreed to hold hands if the fire forced them to make the leap. Reid — his back singed from his trip down to the fourteenth floor to get Cary — had silently given up hope of being rescued; he just wasn't saying so. He had helped throw the drapes out the window to keep them from catching fire, as well as all the carpet that could be pulled up. Now he went to a window, pulled his wallet from his pocket, and took out five twenty-dollar bills. One by one he wadded them up and dropped them, moving from window to window so that the five bills would land at the feet of five different people. He thought some people deserved to be lucky that morning. He wished he were on the street with them.

Horne then pulled his two gasoline courtesy cards out of his wallet and threw them out the window, too. The cards had his name on them, and if he and Cary died, he wanted his family to know it.

Cary, an optimist, wasn't thinking about dying, but she may have been the only one in the room who wasn't. The mood was increasingly grim as the fire kept burning, no rescue arrived at the door, and the stark reality of the situation deepened. Only the fact that the smoke in the suite was tolerable prevented panic. Maude Whiteman had sealed the room tightly at the fire's outset. Wet cloth had been stuffed in every opening, even the keyhole.

When firemen appeared on the roof of the six-story Davison-Paxon's building across Ellis Street and began shooting a deluge gun toward the Winecoff, the nine people in 1612 suddenly felt their spirits lift. Reid had been trying to point the firemen toward Davison's all night.

A brigade had been formed by fireman Walter McDuffie, the brother of Nell McDuffie, the Winecoff executive secretary who had crawled off a ladder into the Mortgage Guarantee Building seconds before Ed Kiker Williams landed on it. McDuffie had run into the alley as soon as he arrived. He saw a woman's body lying face down and was certain it was his sister's. When he realized it wasn't, he ran into the hotel and up the stairs. Three times he tried to reach Nell's eleventh-floor room but collapsed each time. When the night watchman refused to let him into Davison's, he and the other firemen forced their way in.

Fed by three hoses and based on a tripod, the deluge gun produced

With most other guests rescued or dead, the searchlights moved to the nine people still trapped in Room 1612.

1,000 gallons per minute, knocking the glass out of the Winecoff's blackened windows. Two hoses were connected to Davison's standpipes and directed toward guests still alive on that side of the building. They sprayed Gordon Allison, the Winecoff's house doctor, and knocked back another man who was about to jump.

By 6:04 a.m. — two hours and twenty-two minutes after the alarm was sounded — the fire was out. At that moment Capt. Albert Fain heard one of his chiefs say, "Captain, get that roof." Fain, who had been fighting the fire on the Winecoff's middle floors, took one of his men and began trotting up the stairs. He knew he was risking his life, since he was not wearing a mask and his flashlight was out. The men climbed the stairs in darkness, but they made it to the sixteenth floor and found the doorway to the roof across from 1624. They climbed one more flight and then pushed open the door. The fresh air poured in as the two men propped the door open to vent the gases and smoke in the building. When they glanced around, they saw a small cluster of upper-floor guests huddled together, still scared to death. Fain told them to be patient, that the fire was out.

It was 6:30 before the nine people in Suite 1608-10-12 heard a knock on the door. They couldn't believe it when a fireman stumbled in. He said he had come in through the window from the other building and that more firemen were on their way to help them out of the hotel.

"We're going to get you outta here," he said.

But the fireman, struggling to breathe, looked half-dead himself.

"Man, by the looks of you, we're not going to get out of here alive," Reid told him.

Two floors below, Bob and Pauline Bault and Bond and Dorothy Phillips heard a knock on the wall out in the hall. Several minutes earlier, the firemen on Peachtree had shined their big lights into the room, saw that four people were still alive, and began pumping water through the windows. The water — now two inches deep — had set off an electric current in the room, and the four were standing on the bed to keep from being shocked.

When they heard the knocking, the Baults and Phillipses quickly pulled the water-soaked mattress away from the door, and the firemen walked in. All that was left of the door was the frame.

"We're going down now," said one of the firemen.

By then another fireman had made it up to 1612. After another ten minutes, three more came in. The battle to live was over.

CHAPTER 7

B y the time the firemen made it to the top floor, they were bone-weary and wet, their lungs full of black gunk. With the fire out, they faced another task: getting the guests out, both the dead and the living. The living got to go first.

The Hornes told the firemen they didn't need any help getting down the stairs, but they did want to stop by their room, 1606. As soon as they pushed the door open, they could see a body lying on their unmade bed. Cary tried to look away, but she couldn't take her eyes off the man's shoes — curled grotesquely on his feet, the soles nearly bent double from the heat of the floor. Cary also saw that the man was young, in his mid-twenties at the most, and wore a college fraternity ring.

Cary put her shoes on. Reid, who had been sock-footed on the ledge, pulled on his forty-dollar Florsheims. They then left their room for the second time that morning. The body on the bed would follow them out later.

The young man would be identified as James R. Little of Charlotte,

North Carolina. He had been living in 1514, above Ellis Street, for five weeks.

Little had grown up in Atlanta but had moved to Charlotte as a teenager. Three years later he transferred to Georgia Tech, joined the Naval Reserve unit there, and graduated in 1944 with a commission. Assigned to a destroyer, he was involved in seven major sea engagements, winning eleven battle stars and a presidential citation. After the war he took a sales job with Firestone Tire & Rubber in Charlotte. In early November the company sent him to a sales training course in Atlanta, where he had been able to see his brother and several old friends. One of them had invited him to spend Friday night in his home, but Little had declined, saying he had work to do Saturday morning. He returned to the Winecoff around 1:30 a.m.

Little had left his room hoping to find a fire escape or at least a way to the roof. The sixteenth floor was as far as he got. Gasping for breath, he knocked open the first door he came to at the top of the stairs, slammed it shut behind him, and collapsed on the bed.

Despite their protests, the Hornes were accompanied down the stairs by a fireman from Marietta and a sailor who had come up from the bus station to help out. Three floors down, they realized they couldn't have made it without help. They began to realize fully what they had just survived.

The Hornes were seeing the ravages of the fire inside the building, and the vision was harrowing. The bodies in the halls of the sixteenth floor were only the start. By the time they reached the twelfth floor, the walls were scorched black skeletons. Smoke burned their eyes and left them half-choked. Drops of steaming water bounced off their shoulders. The stairway was so dark that they repeatedly called to each other to make sure they didn't become separated. More than a dozen thick hoses wound their way up the stairs, forcing them to search with each step for a safe place to put their feet. Water ran like a swollen creek down the stairs. In the hotel's front doorway, two rows of solemn firefighters, lined shoulder to shoulder, swept the water cascading into the lobby out onto the sidewalk.

It was around seven o'clock — nearly daybreak — when the

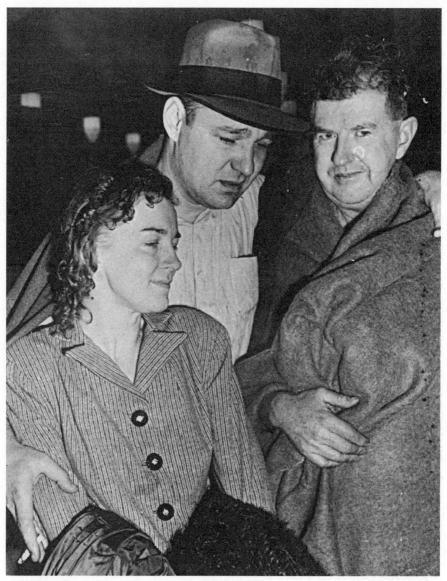

Cary and Reid Horne and J. W. Henry (left to right) reached the safety of Peachtree Street at dawn.

Hornes walked onto Peachtree Street. The fresh air was stunning.

Ambulances waited by the door. Clothes lay strewn up and down the sidewalk and street, thrown from windows by panicking guests. The police were stringing rope from one light pole to another to keep out the curious. Although the fire was out, the crowd was growing, waiting to see the survivors emerge — and the victims. Early risers all over the city had turned on the radio, heard the news, and immediately headed downtown. They stared upward at the fifteen-story brownstone building, standing as firmly as the day it was completed and pronounced fireproof.

The pitiful shreds of sheet, white against blackened window frames, told the story of the morning. Lifelines of hope in the dark of night, they now hung limply, only occasionally twisting in the wind, reminders more of tragedy than of survival. Curtains fluttered in paneless windows. Wisps of smoke escaped into the chilly morning air.

The sun was rising on a nightmare. And if the shocked silence and stark sights of morning were not enough, a sickening odor enveloped the entire block, one that most of the spectators had never smelled — the awful odor of burned human flesh.

The first firemen to arrive at the hotel were now being allowed to leave. They made their way across Peachtree to the Holsum Cafeteria, hoping food would chase the burning from their throats. Grimy and black from smoke, they boarded trolleys on Peachtree Street looking more like strange creatures from another land than the heroes they were. Clean-faced firemen from communities outside Atlanta — Marietta, College Park, Decatur, even Fort McPherson — were replacing the weary. Everywhere faces were grim; no unnecessary words were spoken. Inside, Chief Styron remained in charge, still wearing his pajamas under his fire coat.

The firemen were moving through the dark shell of what had been a hotel when its guests had gone to bed. They saw signs of the night before in the streaky light of dawn: a plate of steak bones in one room, soggy toast in another. Imported cognac. Cheap wine. Christmas presents that would never be given.

It was obvious to the firemen that the fire had started on the third floor. Below that — on the mezzanine and in the lobby — there was no sign of flames, but above that the damage was progressively worse on each floor. On the seventh floor, the firemen saw total burnout. In some

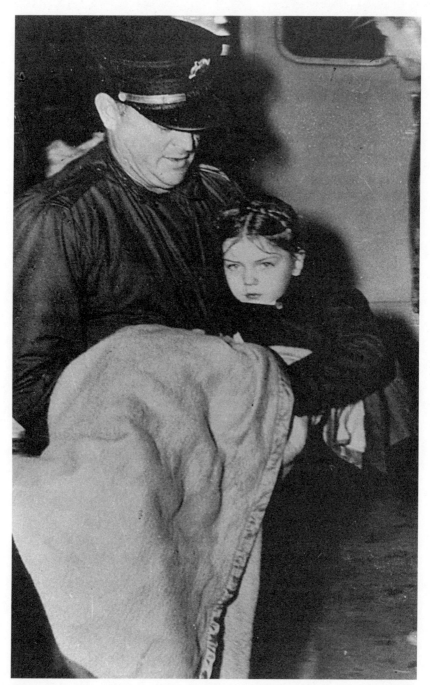

Four-year-old Esther Geele, daughter of the Winecoff's general manager, was carried out of the building by a fireman after being trapped in 1612 with the Hornes.

rooms every stick of wooden furniture had been consumed. Only blackened, twisted bedsprings, dotted with charred bits of mattress, remained. The scene was the same on the eighth and ninth floors, where the firemen believed the flames had reached their zenith. But the tenth, eleventh, and twelfth floors were nearly as devastated.

Big flakes of plaster swam in the puddles on the floor. Walls had disintegrated down to the terra cotta tile, and even that was cracked. Ceilings were scorched and blistered. Charcoal lay inches thick on the floors. Glass shattered by the heat sparkled dully below the windows. Light bulbs had melted, cooling into stalactites as they dripped from their sockets. Telephones were globs of black. Wooden wall radios were gone, as were the wooden blades of ceiling fans. If the closet door was still standing, the knob was a melted mass. The clothes, still on hangers, were charcoal. Toilets were cracked, the water in their narrow necks having turned to steam and ruptured the porcelain.

To the firemen, the room-to-room search seemed almost an invasion of privacy, for each room was now a museum to its former guests and the frightening moments they had spent there. Overturned chairs. Scattered luggage. Beds dragged to the window so a sheet rope could be tied to them. Small personal items, valuable only to their owner, lying on charred dresser tops. A burned-out crib. Toy soldiers covered in soot. Clothes everywhere. Underwear. Socks and shoes. Entire contents of drawers littered the floor, remnants of a frantic search for a valuable tucked away for the night.

The rooms were deathly quiet. Only the boots of the firemen — crunching on charcoal or sloshing through water — broke the eerie silence. Actually, chaos reigned. No one had figured out a way to designate the rooms of the victims, many of them burned beyond recognition.

The bodies were coming down on hand-held stretchers — at first uncovered. Occasionally the carriers cursed the darkness or the water, but mostly they said nothing, in shock from what they were seeing. Talk was minimal.

"Give me a hand here." "Stretchers coming out." "Make way for the stretchers."

This front hall was typical of the damage on the middle floors.

Rescue workers, like this soldier, had to navigate down the Winecoff's narrow labyrinth of stairs with stretchers.

It was hard labor, made even more gruesome by the smell that permeated the building. Four men tried to bring one 250-pound victim down the labyrinth of stairs on a stretcher. They had to stop at each floor to be relieved by a fresh crew.

On the roof of the Mortgage Guarantee Building, bodies were coming across the alley in long wire body-baskets, guided by firemen leaning out of the Winecoff's fifteenth-floor windows. Unshod feet, white and still, protruded from light-colored blankets covering the face of the dead. A small bundle came across, and someone whispered, "Another child."

Nearly ninety bodies had to be removed from the building, and the stretcher-bearers would often turn away from the rooms of victims to search for the living, sickened by the scenes they saw. Most victims on the lower and middle floors had burned after suffocation. A worker

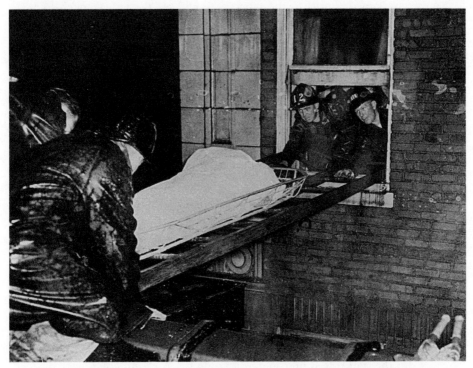

Wire body-baskets from World War II were used to move victims on the upper floors over to the roof of the Mortgage Guarantee Building.

would lift a body at the ankles, and the blackened flesh would simply fall away. When the firemen found whiskey in the rooms, they drank it for courage.

Red Pittman and G. T. Murdock came upon a body that was so hot they had to wash it down with a hose before they could move it. It was only a torso, its head gone and pelvic area obliterated. Pittman opened one door and found an attractive young woman in a green dress sitting by the window, her eyes open and staring outside, as if waiting to be rescued. She was dead. Pittman also found a victim straddling a window sill on the back side, frozen in indecision: jump or try to hold on a little longer.

Fireman W. H. Rogers counted fourteen suffocation victims on the fourteenth floor, the same floor where he discovered a canary chirping. Throughout the hotel bodies were found on beds with smiles seeming to crease their faces, tempting the rescue workers to believe they had never awakened. But then the workers would see the long, scratch-like marks on the walls, as if the room's frantic guests had tried to climb them.

Outside, the wail of sirens had subsided for the first time in more than three hours. All the injured, even the ambulatory survivors, had been sped away, most of them to Grady Hospital ten blocks away.

Since four o'clock an endless parade of ambulances, summoned from hospitals and military bases, had roared and screeched through downtown Atlanta. With each trip, Charlie Brown saw more and more civilian volunteers helping unload the vehicles.

Brown, one of the first ambulance drivers to arrive at the Winecoff, had made so many trips that his siren finally quit. During one stop at Grady, he called a filling station and asked for five gallons of gasoline. By the time he returned from another run, the gas was waiting for him.

Grady's doctors and nurses knew they were facing a major disaster but had no way of gauging the enormity of the fire or number of casualties. With each wave of dead and injured, they wondered if that might be all. Then more would arrive. Then more still.

But the staff was as prepared as it could be. A recent emergency disaster drill had reminded everyone of their stations and duties. Grady's

Ambulances lined up on Peachtree Street at daybreak. Although the Winecoff's awning was in tatters, the Tuberculosis Association's Christmas Seal banner held firm, a symbol of life in a scene of death.

chief resident, Charles Holloway, had reacted quickly after receiving a call from the hospital's executive director, Frank Wilson, who had been alerted by the fire department. Knowing he would need more bed space, Holloway instructed the nurses to rouse patients who could sit up and to call their families to come pick them up. He dismissed anyone who was even slightly ambulatory.

A dozen doctors met the first wave of casualties that came in before five o'clock — the broken bodies of those who had jumped or fallen to the pavement. Three dozen were on hand for the second, larger onslaught. But Holloway and his staff quickly realized they were not going to be able to do all they had hoped; most of the incoming were already dead.

They were wheeled down a gray corridor to what was normally the outpatient clinic and placed on a table or on the floor. The carts went back to the ambulances; there was no time to change the sheets, which were blackened and spotted with blood.

Grady's city morgue had long been filled, for it could hold only twenty bodies. The tables in the bone clinic also filled quickly, and the victims were being sent across Butler Street to the basement of the Grady's "colored" wing, where a temporary morgue was being set up for families to look for their dead. Bodies were everywhere. Some lay on the cold floor, others in the same baskets in which they had been removed from the Winecoff, and still others on top of one another.

Grady's surgeons were ignoring the custom of segregation and working out of both the "white" and "colored" hospitals. But most of the injured lined the halls outside the bone clinic, lying alongside the victims. The doctors' first job was to judge the quick from the dead.

The injured guests now coming in were generally suffering from smoke inhalation or burns, and nurses were watching them for pulmonary edema, a swelling of the lungs. The pace was frantic yet determined, the staff talking in hurried, clipped tones. Volunteer technicians had already come in from other hospitals, as had numerous doctors. The Red Cross began arriving at daylight. Grady had become the sight of Atlanta's greatest medical emergency since September 1, 1864, the day Confederate dead and wounded from the fighting at Jonesboro covered nearly a square block in the center of town — a scene immortalized by *Gone with the Wind*. It was here that the remaining battles between life and death would be fought, here that families would sob in absolute grief.

The survivors sent to Grady as a precaution or to receive treatment for minor injuries were silent, many in shock from all they had seen and from their own near death. But if the dead and injured were quiet, the grieving were not. They cried uncontrollably, and their numbers multiplied quickly.

Atlantans and residents of outlying towns had started arriving at Grady at dawn. Now out-of-town families were showing up at the hospital, trying to discover whether their loved ones were dead or alive. Relatives and friends gathered at the information desk of the emergency clinic and in the waiting room near the main entrance. Grady's beleaguered clerks and secretaries — already overwhelmed by the paperwork of incoming patients — were hardly prepared to deal with the families' questions as well. They had only a partial list of the injured, and there was no list of the dead since most arrived without identification. So they said the same words over and over: "Don't give up hope. This isn't a complete list."

The chaos that surrounded the removal of bodies from the hotel was now creating massive problems. Undertakers had grabbed bodies as fast as they came out of the hotel; a woman's body might have been sent one place and her husband's somewhere else. Relatives arriving at Grady would question anyone they could find: reporters, nurses, other visitors. "He had a bridge," a man said, baring his own teeth to show where the bridge was. A woman in a fur coat told an intern, "She's my aunt. She's about sixty years old and her hair is turning gray." Very few of the imperiled guests had retained the presence of mind of Ashley John Burns, the twenty-six-year-old grandson of Walter Burns, founder of the Burns Detective Agency. Before suffocating in Room 1464, Burns had affixed his Army discharge pin to the lapel of his pajamas.

Even the patients were asking about the missing. Delilah Chambers talked about two-year-old Bob Cox so much that the nurses finally led her to him. Chambers, the baby-sitter charged with caring for Little Bob while his parents visited Atlanta, had seen him bounce on the awning after being tossed out a tenth-floor window by his father but had not seen him be caught by a bystander. The little boy, now orphaned, had been crying continuously at Grady. Overwhelmed with relief to find him alive, Chambers took him into her arms and quieted him.

The condition of the injured varied greatly. Sammie Sue Leadbetter, who jumped from the seventh floor and landed in a net, had only a cracked sternum and burned ankles. She was placed in a room with University of Alabama student Fredna Green, who had passed out on the window ledge of 828 and was still unconscious. She came to after receiving four pints of blood plasma shipped in by the Red Cross. She had minor burns and cuts all over her body but still did not know how she had escaped from the building.

W. F. Robinson, who had also been on the back side of the eighth floor, was one of the few survivors seriously burned. He received third-degree burns on his back, shoulders, and neck when Room 820 turned into an inferno as he awaited rescue on the window ledge.

About a dozen firemen were hospitalized, primarily for smoke inhalation. The most seriously injured was A. J. Burnham, whose back was broken when he was knocked from his ladder by the body of a falling woman. Policeman Bruce White broke his back when he was struck by a teenage girl leaping for a net he was holding in the alley. A part-time professional wrestler, he continued to assist in rescue operations until the fire was out.

⋆⋅ ⊨✦≡ ⋅⋆

The injured tried to notify their families as best they could. Kitty Tribble sent a telegram to her husband in Rockford, Illinois: "Safe after leaping from eighth floor. Was in hotel which caught fire. Thank God."

But many families were already en route to Atlanta. Smith and Janie Dansby from Columbus set out for Atlanta immediately after rising at five o'clock and hearing about the fire on the radio. The Dansbys ventured into the long ward where the injured lay, but their daughter, Bess, was so black from smoke they didn't recognize her. She had sat on the windowsill of Room 730 until she passed out and fell to the court area outside of 330. Her thigh was crushed, and she had nearly bled to death from a deep laceration on her arm, but she would live. It would be days before she was told of the death of her best friend, Dot Tyner, a Youth Assembly observer who had come along to be with Bess. Her body was burned beyond recognition and would be identified by a ring.

The story was the same all over the hotel. One person had lived while a roommate — a spouse, a child, a parent, a friend — had died. Charlie Boschung lived; his wife, Millie, died. Lena Smith lived; her daughter, Anne, fell fifteen stories. Catherine McLaughlin made it down a sheet rope; her sister, Sara Miller, didn't. Rudolph Ogilvie lived; his roommate, Bobby Sollenberger, died. Mary Helen Rasmussen survived the smoke in 1620; her husband and son didn't. Bob Irvin slept in 806 and lived; his brother, Harold, slept in 830 and died.

Usually on Saturday mornings Bess Dansby and several of her friends did a radio show back in Columbus. This Saturday the show was broadcasting that Bess had died — about the same time the Youth Assembly delegates fortunate enough not to be assigned to the Winecoff were filing solemnly into Georgia's gold-domed capitol, three blocks from Grady. They were not the same carefree teenagers who had traveled to Atlanta the day before. Stunned, they talked quietly among themselves or sat silently.

A few minutes before ten, as the teenagers took their seats behind the large wooden desks of the House of Representatives, a hush settled over the room. The YMCA leaders, entrusted with the caring of so many children, spoke in voices choked with emotion. They began to call the roll of the 316 boys and girls. Each time silence followed the name of one of the forty who had stayed at the Winecoff, the teenagers glanced at each other in somber puzzlement. What did that mean? Were they dead? Injured? Unaccounted for?

The assembly was scheduled to continue through the day and conclude with a banquet that night at the Ansley and a closing session Sunday morning. Instead, after the roll call, state YMCA secretary Hugh Robinson tearfully sent the delegates home to their families. A prayer was offered.

Three delegates — Buddy Elliott, Earle May, and Ramsey Simmons — spent the morning at Grady searching for their Bainbridge High School classmates. When told to send in their Youth Assembly registration forms — with Winecoff marked as the hotel preference — the three boys had delayed so as not to have to stay with the rest of the delegation. Buddy and Earle were seniors and Ramsey a junior, and they were ready to be on their own, away from the watchful eye of their chaperon, Mary Alice Davis.

The boys were assigned to the Piedmont, while the other seven Bainbridge students and Miss Davis stayed at the Winecoff. Friday evening Buddy and Ramsey took two of the Bainbridge girls, Maxine

Willis and Sue Broome, out to a movie and to get a Coke. They returned to the Winecoff with the girls and left shortly after midnight. When the boys heard sirens during the night, they assumed it was only the noise of the city, turned over, and went back to sleep.

At Grady, they could find neither Maxine nor Sue. But Earle was able to identify Clarence Bates — from an ID bracelet he had noticed during the five-hour ride to Atlanta. Ramsey identified the body of Mary Lou Murphy.

The hospital clerks cautioned the families of the victims that not all the injured had come to Grady, that some were at other hospitals. But rather than going to Crawford Long or Georgia Baptist or Piedmont, most families crossed Butler Street to stand in the growing line on the sidewalk. Many were already weeping, or talking quietly about a ring or a necklace or maybe even a filling that could be used to identify a burned body.

Relatives and friends passed through the makeshift morgue at a rate of fifty an hour. Interns accompanied the searchers through the room, viewing the same corpses again and again — a chilling job. About every ten minutes a body would be identified, and the weeping and sobbing would begin anew. A clerk at a desk would type up an identification tag to be attached to the victim's toes, and undertakers waiting outside would come in, zip the body into a black rubber body bag, and take it out, passing through the crowd of families checking the latest lists of the dead and injured.

Courage — or sheer desperation — was required to enter the room. A ruddy-faced man who had spent his life in the sun hesitated briefly at the doorway before reaching down to grab the hand of a younger man. An elderly man, his face swollen, clung to the shoulder of a friend as he peered at a tattoo of Air Corps wings on the arm of a fire-blackened body. He shook his head, almost in happiness. "No," he said, "that's not him." Another man refused to confirm the obvious identification of his son, unable to confront his death.

Atlantans searching for relatives of friends from out of town would take notes and then go to the phone, call the family, and ask, "Well, what was she wearing?" One woman emerged from the room asking herself

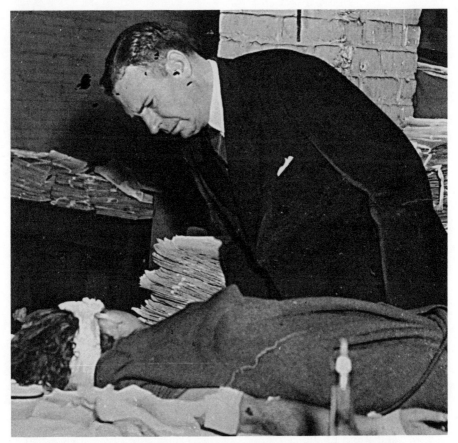

Georgia adjutant general Marvin Griffin searches at Grady Hospital for his fourteen-year-old daughter, Patsy, a Youth Assembly delegate who was killed in the fire. Eight years later, Griffin would become governor of Georgia.

over and over, "How am I going to tell that mother that her child is dead?"

Relief occasionally broke the grim pall. A woman came out of the makeshift morgue and, with a lift in her voice, called to a friend, "No, she isn't in there." An intern then gave the woman a list of mortuaries to check.

Atlanta was responding to its tragedy in Herculean fashion. Even before daybreak, guests who had escaped from the Winecoff uninjured were embraced by onlookers and taken to hotel rooms or into homes.

Blood donations were called for over the radio, but by ten o'clock the line of donors at Grady was so long that the call was canceled.

The Red Cross mobilized a volunteer force of 250 people. Nurses were sent into hospitals, while others were sent to Atlanta's bus and train stations to meet the families of the dead and injured. Still others answered the phone. Fourteen lines set up at the Red Cross headquarters on Peachtree Street would take in an estimated 10,000 calls Saturday.

The world wanted to learn more about Atlanta's horrible fire. Every long-distance operator in Atlanta was on duty. Trying to meet their Sunday deadlines, London newspapers called continuously. So did Americans, frantic for news of family and friends. For the Winecoff and its guests, the Salvation Army acted as intercessor, fielding all calls — and even telegrams — at its Atlanta headquarters a block away from the hotel. The Western Union office at the corner of Forsyth and Marietta was swamped.

At the Winecoff itself, a parade of automobiles started as soon as the fire trucks pulled away. The cars would circle the block and come down Peachtree again for a second look, and then a third and a fourth. Around and around they'd go, with passengers leaning out of windows and craning to see fragments of sheets hanging from the windows. By midday, more and more of the cars bore out-of-town license plates. Police had blocked off the sidewalks around the hotel to prevent looting, but on the sidewalks across Peachtree and Ellis streets people stood four and five deep, talking to their companions and shaking their heads.

Far from the crowds at the Winecoff, the families of its former guests were driving all over the city in a fog of apprehension, grief, and desperate hope. Many of them had been half-asleep when they heard about the fire. The trip to Atlanta, with its wrenching uncertainty, had been an emotional nightmare. And once they were in Atlanta, the pain of not knowing only intensified. If the family member wasn't among the injured at Grady or another hospital, then the search turned to the bodies. Lift a sheet and look into the face of death. Put the sheet back in place and move on to the next corpse. The search stopped only when a sheet was lifted and a son or daughter — or father or mother, brother or

Throughout the day the curious flocked to Peachtree Street, which was made even more congested by trolley tracks that ran up and down the middle of the street.

sister — was underneath.

When the parents of Harriett Strickland and Jerry Jenkins arrived from Donalsonville to search for their daughters, they were accompanied to Grady by Gloria Daniels, one of the two Donalsonville freshmen who had been scheduled to stay at the Winecoff but had been sent to the Henry Grady instead. The six Donalsonville girls had discussed seeing *Song of the South* and then all spending the night together in Room 930. But when Gloria and Zula Marie Jones stopped by the Winecoff on the way to the movie, the older girls weren't there. The two freshmen went on to the movie by themselves and, when it was over, headed back to the Henry Grady.

Only twelve years old, Gloria didn't know how to tell the Stricklands that she had already been to Grady and seen Harriett's body there. So she went back to the hospital with the Stricklands, where they identified their daughter's body themselves. Gloria then went with the Jenkinses from funeral home to funeral home, searching for their daughter, Jerry. They found her other two roommates, Robbie June Moye and Jeannette Riley, and then, sadly, found their own daughter.

A five-person search party found the body of Nell Sims Saturday afternoon at Hemperley's. Nell had fallen fourteen stories while trying to follow the Hornes across the ledge below the fifteenth-floor windows. Her family was notified by Reeves Marquis, the friend Nell had come to Atlanta to go shopping with. Reeves was still with her boyfriend on one of the lower floors when the fire broke out. As soon as she escaped, she called her mother, Sadie, who called the Sims' house and told them of the fire.

Sadie Marquis drove Ida Mae and Martha Sims to Atlanta. They got a room at the Piedmont, which became the headquarters for their search. Reeves was there and so was Nell's brother, Elrod, who had come from Athens. It was an awkward situation. They all wanted to find Nell, but the Sims' bitterness toward Reeves was difficult to conceal. "Reeves is a more experienced person than Nell," Martha, Nell's sister, told her mother. "If she had been with her, maybe they could have figured something out."

It was Reeves who finally found Nell's body at Hemperley's.

Christy Hinson's father burned up a brand new DeSoto on Highway 19 between Thomaston and Atlanta. He drove it right out the front door of his dealership, with his wife in the front seat and his son in the back. They reached the Winecoff just as the firemen were extinguishing the flames. They stood across Peachtree Street and gazed up at the building, so close yet so far from Christy, a seventeen-year-old senior at R. E. Lee Institute.

Unable to find her among the bodies at Grady, the Hinsons drove from mortuary to mortuary, listening to WSB Radio's constantly updated reports on where bodies could be found. At mid-morning the station reported that Hemperley Funeral Home in East Point had the body of a girl wearing a 1947 R. E. Lee ring with the initials C.A.H. inside. But the Hinsons heard only the last part of the report, not the location of the body, so they continued their trek through funeral homes, having to look at body after body. Finally, in the middle of the afternoon, the Red Cross told them Christy was at Hemperley's. Four Thomaston seniors had suffocated in 1430, yet their bodies had been taken to three different locations.

Several badly burned bodies — including those of two twenty-one-year-old friends, Freddie Louise Pruitt and Emelda Reeves — were identified by a special investigative team assembled by the Salvation Army. Neither woman was registered at the hotel, but Pruitt was clutching a high school ring from Lenoir City, Tennessee, tightly in her hand. She was identified through the "F.L.P." engraved on the ring. Because the other body was found nearby, the Pruitt family was asked about any close friends their daughter might have. Emelda Reeves' name was mentioned, and Reeves' mother identified her through a peculiar tooth they had discussed having removed.

Pruitt had moved to Atlanta two years earlier because of better job opportunities and had found work as a secretary in a downtown insurance office. Reeves worked for a wholesale liquor distributor. She would be buried by the Salvation Army.

Families who lived far away heard of their loved ones' deaths in a variety of ways. Dorothy McDonald was grocery shopping in Memphis Saturday morning when a friend mentioned that there had been a terrible fire in Atlanta the night before. Afraid to ask where the fire had been, Dorothy hurried home. A call soon told her that her husband, Jimmy, was dead. J. O. Godbout of Long Island, New York, learned of the death of his twenty-three-year-old son-in-law, John Irwin, while reading the Winecoff casualty list in a New York paper.

Andrew and Christina Conzett of Dubuque, Iowa, were called by a newspaper reporter and asked if they had been notified of their oldest son's death. They couldn't believe it; he had survived too much to be killed in a hotel fire.

A lanky man of few words, Lt. Cmdr. Elmer "Spike" Conzett was a war hero of such stature that he was written about in *Guadalcanal Diary.* In World War II's first major naval battle — Midway — his aircraft carrier, *The Yorktown,* was sunk, and he was plucked from the Pacific. Later that summer of '42, Conzett and another dive-bomber pilot took their planes to their customary 10,000 feet, put their sights on a Japanese destroyer escorting landing ships into Guadalcanal, and dove. Conzett's cockpit was shot up by anti-aircraft fire, and a piece of shrapnel from a 200-mm shell went through his leg and into his shin. His instrument panel was gone, but he finished his dive, dropped his 500-pound bomb, and sank the destroyer. After recovering in San Diego, he returned to the Pacific and in 1944 participated in the bombing of Truk, where twenty-three enemy ships were sunk.

Now, on the fifth anniversary of Pearl Harbor, the attack that he risked his life to avenge, he had died back home in America. It was more than his parents could take. Ironically, that same day, they received a letter addressed to Spike on Winecoff stationery. It was from his girlfriend, who had spent the night in the hotel during the middle of the week and had hoped to see Conzett there. He had known he would be making a stopover sometime during the week but had not arrived until Friday, when he and one of his men, Ens. R. C. Williams, were assigned Room 1230. It didn't matter that Conzett, disoriented from loss of blood and flying without a control panel, had been able to pull out of a near-disastrous spin and find his way back to Guadalcanal in semidarkness and a soup-like cloud cover; in 1230 he never had a chance. Even the most cursory inspection of the Winecoff on Saturday would show the "30" rooms to be the most devastated. They were black and deep in ash, looking like bombed-out hulls on some floors.

Guests trickled in all day trying to retrieve their belongings, if anything was left of them. One man took a look at the inside of the building and became hysterical. Kelly Hammond, who escaped from 1018 with his wife and the Turk brothers, returned to his room and saw the ravages of the fire. The steel frame of the bed had collapsed in the middle and the springs had melted. Hammond's palms sweated all day.

At dusk, water still dripped from the lobby chandelier into a washtub full of discarded paper cups. Reams of ruined stationery lay scattered across the floor, all of it describing the Winecoff in the same words: "Absolutely Fireproof." Postcards home, left at the desk by guests Friday evening, were

Onlookers in front of Davison-Paxon's on Ellis Street were restrained by roping intended to prevent looting.

still awaiting pickup. "We've finished our shopping," said one. "See you Sunday."

CHAPTER 8

⊶ ⚔ ⊷

The story was buried on Page 12-A of Sunday's Atlanta *Constitution*. Among the fifty-five articles on the Winecoff fire in Atlanta's two Sunday papers, perhaps no other so aptly summed up the pain and futility. The headline read simply: "Bodies Described." Then followed a description of the eighteen bodies still not identified Saturday night:

Child. Boy. Brown overall. 4-6. New brown shoes. . . .

Female charred. About 16-25 years old. Long dark brown hair. Identification bracelet with name "Jean". . . .

Charred male. Medium height.

Female, 55-65 years old. Long gray hair. Hotel room key 1402 in hand. . . .

Man. Navy type black shoes, 9 1\2. Gray pajamas, yellow trim. Light brown unruly hair. . . .

Baby girl. About 18 months-2 years . . . Blonde hair. Blue eyes.

Charred female. Brown long hair. Charred beyond an

identification.

Female, gray hair, young face . . . Old abdominal midline incision.

And so it went, from new brown shoes to an old abdominal incision, each short paragraph more horrifying than entire name-filled articles.

Also in Sunday's *Constitution*, editor Ralph McGill, having toured the building with other reporters, wrote a front-page story that captured the fire's incredible horror:

Down there on the 12th floor is a woman sitting in the window. Her arm is on the ledge. A flashlight in the dim gray of first dawn shows the skin clean and white, the nails scarlet with enamel. But where is her head? It is gone, burned away by the blowtorch heat of exploding burning gases. . . .

I recall how incongruous was the woman with her head blown off. She was fully clothed, even to her coat. In the hand on the window ledge she held a pocketbook. When they went to move her body, they had to literally tear the pocketbook from the stiffened fingers.

"The clutch of death," said a fireman. . . .

Like a scene out of Dante's *Inferno* I saw this. In the blaze of a portable searchlight came two stretcher-bearers.

The body of a man was covered with a pink, wet blanket.

As they walked, stumbling, his head, blanket-wrapped, shook negatively from side to side.

"No, no, no, no, no," his head kept saying as they walked to the stairway and down it. "No, no, no, God, this should not have happened to me. I have a wife and children at home, God. They need me." Or, "I have a sweetheart. I am to marry her next week. No, no, no, no, God, don't let this be me. . . ."

You may smell the sweet smell of new death. It is something like going to a barbecue but it is a horrible sort of barbecue. You hate the thought; it comes nevertheless. It smells of scorched flesh. . . .

Sunday morning, prayers were raised for the fire's victims in churches across the city. Herman Turner, pastor of the Covenant Presbyterian Church, asked for "God's grace for a city and state bogged down in this grief."

Meanwhile, sixty-four people — one-third of all the fire's survivors — remained hospitalized, eleven in critical condition. Their families continued to flock to the city during the day, just as the parade of cars continued to circle the Winecoff. Inside, electricians had installed a temporary lighting system that was allowing a stream of officials to inspect the building. Guests and their relatives came by to claim belongings, but they weren't allowed above the lobby without a Burns detective or city policeman. In most cases they were told they couldn't go get their things until Tuesday — "mayor's orders."

William B. Hartsfield, in his eighth year as Atlanta's mayor, arrived at the Winecoff early Saturday morning in a topcoat and hat, his face grim and unshaven. He talked with both police chief Marion Hornsby and fire chief Charles Styron. Upon parting, he told Styron, "Come by my office Monday, and tell me what sort of equipment you need. We'll get you anything you want."

After leaving the Winecoff, Hartsfield wondered what the world would think of his beloved Atlanta now. He knew he had a major political problem to solve. The worst hotel fire in history had occurred in his city. Local ordinances had not required the building to have a sprinkler system, fire escapes, or fire doors. On top of that, an eighty-five-foot aerial ladder had been sitting in a repair shop.

Hartsfield realized a thorough investigation of the fire's origin would be necessary and that it might be his administration's biggest undertaking ever. If the fire had been set, Hartsfield knew it would be up to his police department to find the arsonist. He also knew that arson was one of the toughest crimes to solve.

But Hartsfield was an optimist as well as a shrewd politician. He decided to salvage what benefit he could from the fire by making Atlanta's tragedy an American learning experience. He invited every fire prevention group in the country to come and inspect the Winecoff.

"There have been too many such fires in the country and too many in Atlanta itself," Hartsfield told reporters in the Winecoff lobby Sunday,

Mayor William Hartsfield, Fire Chief Charlie Styron and investigators Carey Hutson and H.N. Pye (left to right) gather in the Winecoff lobby on Sunday. Hartsfield would remain mayor for another sixteen years, during which time he would make Atlanta the aviation center of the South and guide the city through the turbulent desegregation era as "the city too busy to hate."

perhaps thinking of the great Atlanta fire of 1917 that left his family homeless when he was a boy. "I am asking the fire prevention agencies around the country to send representatives here. I have given orders that nothing is to be touched for another twenty-four hours in order that they can see for themselves what happened and determine what has to be done to prevent another fire like this one."

⋯ ≼♦≽ ⋯

Late Sunday evening, just before midnight, twenty-year-old Edith Burch of Chattanooga, Tennessee, became the fire's 118th victim. She had amazed Grady's doctors and nurses by living through her fall from a sheet rope, but her injuries were too much to overcome. She had fallen through the hotel's canopy, breaking both legs and arms and suffering

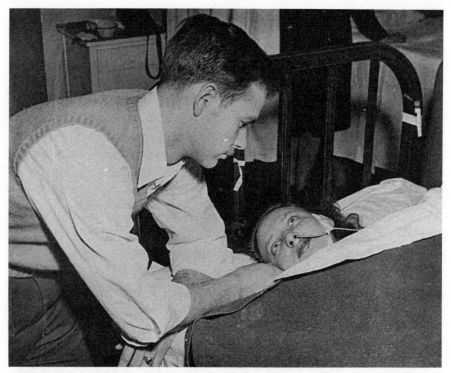

George Burch at Grady Hospital with his twenty-year-old wife, Edith. She died Sunday night from injuries suffered when she fell from a sheet rope.

internal injuries. George Burch, her husband of six months who survived by coming down the sheet rope after she fell, was at her side when she died.

Monday was a day of mourning across Georgia. Funerals were held in churches filled to overflowing. In Thomaston schools closed and flags flew at half-mast in memory of the four senior girls and their chaperon who had suffocated in the fire. Their funerals were intentionally spaced throughout the day — at 9:30, 11:00, 2:00, 3:00, and 4:00 — so friends would be able to attend as many as possible.

In Albany a joint service for three Youth Assembly victims was led by ministers from the First Baptist, First Methodist, and First Presbyterian Churches and drew six truckloads of flowers. The three girls were buried in adjoining graves.

A mass funeral was also held at the Donalsonville cemetery, with the town's four ministers leading the service. The local newspaper referred to the death of four Seminole High School girls as the town's "greatest loss."

Seventeen miles down Highway 84 in Bainbridge, the funerals started Sunday and continued through Tuesday. With the loss of seven high school students and their chaperon, this town was hit harder than any.

Businesses closed in Fitzgerald for the funeral of Dot Smith and her three children. When that service was over, Mamie Kiker and other family members rode over to Cordele for the funeral of Boisclair and Claire Williams. In a single day, Mamie Kiker witnessed the burial of

Dorothy Smith and her three children — Fred, Dotsy, and Mary — were buried Monday in Fitzgerald. Dorothy's sister, Boisclair Williams, and her daughter, Claire, were buried the same day in Cordele.

two daughters and four grandchildren.

Georgians throughout the state read McGill's front-page article Monday. Having written Sunday about the fire's horror, he turned his attention to how such fires could be prevented:

> The phrase "fireproof building" is out of the books of the fire underwriters and it applies only to insurance rates. That is the great and evil fallacy of the word "fireproof." The Winecoff still stands. It did not burn. It is, therefore, "fireproof." . . . It is no comfort to know the building can stand great heat across long hours without structural failure. Human beings can't.

McGill's column referred often to an article in the October issue of *Atlantic Monthly* entitled "What is Fireproof?" and written by Maurice Webster, an architect who served on the investigating jury that studied the La Salle fire in the summer. McGill called him "America's best informed man on hotel fires," and his lengthy excerpt from the article began:

> The story of hotel fires in which many lives are lost follows a pattern which is so repetitious that certain conclusions seem inevitable . . . an area in the ground floor (or near it) bursts into flames so rapidly that fire-fighting equipment cannot prevent the first nearly explosive combustion. The enormous expansion of flaming gases creates its own draft, carrying smoke and fire up open stairways and elevator shafts. Each well or shaft is like a chimney except that the outlet at the top is safely sealed, while holes in its sides lead to bedroom corridors.

The article, published two months before the Winecoff fire, might have been read as a dire warning to hotels such as the Winecoff, but Atlanta had done nothing. McGill, so often the city's conscience, had struck a telling blow. Atlanta looked foolish.

Certainly, Atlanta looked like a dangerous place to visit to the Gainesville families who spent part of Monday in Room 1130 of the

Winecoff, searching for pieces of anything that might help them identify their daughters. The four Gainesville girls who stayed in 1130 were still listed as missing. The only thing found in their room Monday was a pocket-size New Testament and remnants of a Youth Assembly ribbon.

Suzanne Moore and Ella Sue Mitchum were finally identified by a Gainesville dentist, who used dental records and an enlarged photo of the victims taken for their high school annual. Unable to distinguish Frances Thompson and Gwen McCoy by means of the bits of clothing stuck to their charred bodies, their parents eventually decided to bury the two friends together. They shared a tombstone.

Tom Bolton, the British vice consul in Atlanta, identified his mother-in-law, Borgia McCoy, by a piece of nightgown and a few strands of hair. McCoy, fifty-six, had been a buyer of infant clothes for a department store in Portland, Maine, before moving to Atlanta with the Boltons to help care for their baby. In November, as the Boltons prepared to move into their new home in Druid Hills, McCoy returned to Maine to pick up furniture in storage. When she returned to Atlanta Wednesday, December 4, she took a room at the Winecoff for a few days rather than crowd her daughter's family at the Buckhead rooming house where they were staying until their house was completed.

The body of thirty-six-year-old Atlanta beautician Laura Miranda was not identified until Monday simply because no one had been looking for her. Miranda had grown up in a large family, had her first child by sixteen, and was divorced by twenty-one. Unable to support her children financially, she had given them up to her husband's family. She lived with her sister on Highland Avenue and worked at Al Taylor's Beauty Shop on the third floor of the Grand Building, across from the Winecoff, giving permanents to working girls. Saturdays were busy days, so she frequently spent Friday nights in a downtown hotel rather than take the trolley home and then catch it again early the next morning. After Miranda's sister didn't hear from her all weekend, she notified other family members. They found Miranda's badly burned body among those still unclaimed, identifying it through dental records. She had rented Room 826 for the night.

Investigators continued Monday to walk the soggy, blackened halls of the dimly lit Winecoff. City detectives, meanwhile, interrogated hotel employees and used radio appeals to locate guests on the third or fourth floors, where it was assumed the fire had originated. Fulton County solicitor general E. E. "Shorty" Andrews announced that the grand jury would undertake a complete investigation. "If the hotel management is free of fault and negligence, it has the right to have this passed upon in an official investigation," he said. "If there was some oversight, the public has the right to know it."

Monday's biggest event was the five o'clock meeting of the city council's board of firemasters at City Hall, a block from the capitol. Fire marshal Harry Phillips told the board plainly — but prematurely — that the cause of the fire would probably never be established definitely. "We've been unable to locate anyone who saw the flames before several floors were totally engulfed in fire."

Phillips said the investigation pointed to a folding roll-away bed left in the hall near Room 326. He said he had located the two men who stayed in that room — John McFalls and Nolan Russell, the two utility linemen from Andrews, North Carolina, who had come to Atlanta to buy a jeep — and had talked to them by phone earlier in the day. They told him they didn't know how the fire started.

Carey Hutson, an investigator from the National Board of Fire Underwriters, then presented a report also signed by Chief Styron and H. N. Pye, the chief engineer of the Southeastern Underwriters Association. The report stated that the fire started in the partially burned mattress outside 326. Hutson said he was unsure of the fire's cause.

After giving his report, Hutson took questions from the floor. Solicitor General Andrews — whose position was comparable to district attorney elsewhere — asked if negligence by the hotel owners or operators had been noticeable. Hutson replied that he had seen no indication of negligence.

Hutson pointed out that the 1943 Atlanta building code — based on recommendations of the National Board of Fire Underwriters — stated that safety measures could be made retroactive for existing buildings on written order from city officials. But city building inspector Marvin Harper countered by saying the city attorney had ruled that buildings constructed before 1943 could not be forced to make alterations despite the retroactive clause.

Mayor Hartsfield said he didn't want to pass ordinances in the heat of the moment and then have to undo them later. "I don't see why we should go all out with these changes just in Atlanta when there should be changes in other towns too. Why should we make it safe in Atlanta when Atlantans going to other towns would be in the same danger?" asked the mayor. He would prefer, he added, that recommendations by national fire prevention groups be mandated for all cities.

Fire Marshall Phillips then stood up, saying, "What we need now is a yes or no answer to the question: Can we go back to old buildings and require that the owners make them safe?"

Phillips pointed out that eighty percent of the city's buildings did not comply with the code in some manner, but that whenever he had attempted to get the city to mandate fire doors, sprinkler systems, or other safeguards, the attorneys had ruled the city didn't have that authority. He said the city attorney had told him that a building permit — requiring compliance with the laws on the books at that time — was a contract between the city and the property owner and that new codes could not be retroactive.

Hutson spoke up again: "I don't believe any city has a right to give a man a contract to burn up his neighbor. . . . I know other cities have outlawed this form of menace. Some cities have even ordered old buildings torn down when they're deemed firetraps under new codes."

Mayor Hartsfield repeated that he didn't think there should be any rush to do something that might have to be undone later. Always an ally of business, he realized the political repercussions of forcing expensive structural alterations on major property owners.

Alderman James E. Jackson interrupted: "I don't think we would be accused of wild hysteria if we insisted on installation of sprinkler systems and possibly enclosed stairwells."

The mayor responded that such action might be made impractical by legal technicalities — such as cases where buildings were owned by insane people or where the title was in dispute.

"My observation," replied Jackson, "has been that some of the largest businesses rent fireproof buildings and install sprinkler systems to save the contents they have in storage. But here we have a number of hotels whose contents are human beings, and we sit idly by and say they are safe."

Hartsfield asked Fire Marshall Phillips if cigarettes weren't the chief cause of hotel fires. Phillips said yes. The mayor asked if cigarette fires weren't usually caused by guests who had been drinking. Phillips again said yes. Hartsfield then said a lot of whiskey bottles had been found at the Winecoff. He asked if the person chiefly responsible for the fire weren't "John Barleycorn."

<hr />

Tuesday Atlantans learned of the kangaroo-court atmosphere of the fire board meeting. They also learned of the Winecoff's 119th and final victim. Early in the day Miami car salesman Louis Cochran died at Grady Hospital from injuries he suffered in a fall when his sheet rope caught fire.

With 119 victims, approximately forty percent of the Winecoff's more than 280 guests had died. Another twenty percent or so — sixty people — had been injured, meaning only forty percent of the Winecoff's guests survived unscathed, physically at least.

The fire was officially listed as lasting 142 minutes — from the first alarm at 3:42 until the last flames were extinguished at 6:04. So there was an average of nearly one death per minute. Thirty-two guests died from injuries suffered in falls or jumps, forty suffocated and thirty-seven were listed as burn fatalities, although virtually all were believed to have suffocated before the flames reached them. Sixteen people died on the twelfth floor alone. More than eighty percent of the deaths occurred above the eighth floor, out of the reach of aerial ladders.

Also in the *Journal* Tuesday was an eloquent letter from Augusta author Edison Marshall. It hit the city harder than McGill had Monday. Marshall wrote that although the fire had occurred in Atlanta, most of the funerals were being held in towns and cities throughout Georgia. He went on to say:

Atlanta is our capital. We often have business there. Atlanta's stores invite us to shop at the Christmas season. We take our families there for a holiday. . . . We are told we are welcome there and we take for granted that we will be protected from death striking in the night — death by torture. . . .

Atlanta, you permitted, you were party to, this tragic deception. You cannot put all the blame on the hotel's managers and owners and incompetent inspectors. You let the hotel stand on your busiest street; you gave room for its lighted marquee; you taxed and licensed it. . . . Only a month ago your officials inspected it and found it satisfactory. Did your official inspectors look for fire escapes?

Had you afforded us any protection at all? Were there fire doors or sprinkler systems? We do not ask what started the fire but by what great blowtorch did it envelop in a matter of minutes the stairways and elevator shafts, trapping our people in a cage of flame? Why did a young mother have to stand at a window, hurl her two babies to smashing death and follow them screaming? . . . You are a rich, populous, sovereign city but your hotel fires are our business as much as yours.

Atlantans were embarrassed. Their city apparently had been very nonchalant in its attitude toward fire prevention, a fact that was increasingly obvious with each edition of the Atlanta newspapers. One issue reprinted a foreboding column written by Wright Bryan exactly six months before the fire. It ran June 7, two days after the fire at the La Salle, another fireproof hotel:

Suppose a tragedy like Chicago's La Salle Hotel fire struck Atlanta! . . . Would we have a clean civic conscience? Could we charge it off to providential reasons and say that everything humanly possible had been done to prevent such a disaster? Or would we be forced to admit that negligence had destroyed human life? . . .

But if Atlantans wanted something done at least to partially mitigate the horrible deaths of so many innocent people, it was becoming apparent that their city was not going to act quickly, if at all. The only positive step — other than the memorial service scheduled for Sunday by the Atlanta Christian Council at the City Auditorium — was a promise from Alderman Jackson to push for an ordinance requiring hotels to make hourly inspections for fire.

Other southern cities were responding swiftly to Atlanta's fire. Knoxville ordered hotels and apartment houses to install fire escapes or close in seven days. Officials in Birmingham, Montgomery, New Orleans, Little Rock, and Memphis, meanwhile, said they would seek tougher laws to prevent such a fire from occurring in their city. Even legislators as far away as New York and New Jersey said they would push for tougher requirements.

Atlanta's "official" position was perhaps best summed up at the end of a two-and-a-half-minute Paramount newsreel:

> Grief-stricken Atlanta takes prompt steps to invite in experts of the nation's fire prevention agencies, who will study the lessons here. Plans move ahead to insure that another such hotel tragedy may never happen.

On Wednesday, the Winecoff's owner, Annie Lee Irwin, sold the hotel for $570,000 to the Arlington Corporation, which had been incorporated that same day and listed Mrs. Irwin as one of its stockholders. The sale was both curious and foolish, since it seemed to be an admittance by the Irwins that they were culpable and trying to dodge responsibility. If the Irwins were not liable, then there would have been no need for the sale. If they were liable, then the sale would, of course, be set aside. One of the Arlington stockholders admitted in the newspaper that the sale had not relieved Mrs. Irwin of any liability as owner, but the damage had been done. The city had already formed its opinion: the wealthy owner of a hotel in which 119 innocent people had perished was trying to beat a rap.

Both Mrs. Irwin and the hotel's operators — the Geeles and Robert O'Connell — hurt themselves immeasurably by their public silence during the days after the fire. Speaking through her real estate agent, Mrs. Irwin had said only that the hotel would be rebuilt "with every safety device required or recommended." Meanwhile, the Geeles — Art Jr. and Art Sr., who sped back from their deer-hunting vacation in southeast Georgia — declined comment except to say they were covered by insurance.

What should have been the biggest story of the week following the fire didn't break until Saturday, December 14. It received only bottom-of-the-front-page treatment in the Sunday *Journal* and was buried inside the *Constitution*. The *Journal* headline read: "Incendiary Origin in Hotel Fire Suggested by Gotham Expert." It was the first time an investigator had publicly stated the opinion that the blaze was set deliberately. The investigator was John J. McCarthy, a fire prevention consultant for the New York Hotel Association and retired chief of the New York City fire department.

As early as the day after the fire, the Associated Press had reported: "The speed with which Saturday morning's blaze spread has baffled officials." McCarthy arrived in town Sunday night, cased the Winecoff Monday, and issued his opinion five days later: "Either one of two possibilities is tenable. It was either arson, or the fire was burning a long time before discovery. In my thirty-two years of fire-fighting experience, I have never seen such devastating effects from such a small point of origin."

━ ≕✦≔ ━

Although city officials were reluctant to discuss arson publicly, it was a popular topic of conversation in the days immediately after the fire. One early rumor was that the fire had been set to cover up a murder, but the coroner's office reported that none of the victims had been shot. The more ludicrous rumors made the whole arson theory easier to dismiss, however. One of the most absurd was that the Irwins, the hotel owners, paid someone to set the fire so they could collect on their insurance policy. The hotel's primary insurer — Ocean Accident and Guarantee Insurance Corporation — was to be the fire's most thorough and persistent investigator.

Another rumor held that a juvenile pyromaniac had escaped from the state training school six months earlier and contacted his mother by telephone on the day of the fire. He was considered the culprit in two false alarms called in the night before the fire from a call-box near the Winecoff. Two fourteen-year-old boys were indeed questioned by fire marshal Harry Phillips after having been arrested for starting a small fire in the basement of the Mortgage Guarantee Building. They supposedly had been passing through the alley behind the Winecoff and began throwing matches through a hole in the Mortgage Guarantee's rear door and onto a pile of waste paper. But both were able to account for their whereabouts on the night of the Winecoff fire.

The investigators naturally looked into anyone who might have had a grudge against the Winecoff or its owners or operators. They investigated Nick Skartsiris, for example, who six weeks before the fire had become very upset when he was thrown out of Room 514. He had been living at the Winecoff and supposedly buying whiskey from Greek liquor stores and then selling it to guests at inflated prices. Police also briefly investigated a purse snatcher who had been thrown out of the Plantation Room in November.

The Winecoff's clientele offered fertile ground for rumors of revenge. Although the hotel still enjoyed a fine reputation, it also had a seamy underside. It could be the site of an exquisite dinner on fine china, a gambling game, or a clandestine rendezvous. Phony-sounding aliases went unquestioned at the front desk — Mr. and Mrs. Johnny K. Jones from Jonesboro were regulars. The fact that prostitutes were among the fire's casualties only heightened suspicion of arson.

On the weekend of December 21-22 — two weeks after the fire — arson theories dominated the news. On Sunday Edwin Folz, the agent in charge of the Atlanta office of the FBI, said he had given police chief Marion Hornsby a lead. Hornsby said he had been told "the name of a man who could put police in contact with a man who had overheard a third man say he had started the fire." Hornsby said detectives were already checking out the lead and he would report any progress. He never did.

Also Sunday, a fire at the Kentucky Hotel in Louisville prompted the

fire marshal in Dallas, Texas, to say that Texas authorities were considering a possible connection between several arson attempts in Houston
that year and hotel fires elsewhere in the South. The Dallas fire marshal,
B. C. Hilton, had inspected the Winecoff and said he thought it was
arson. No further developments in that investigation were ever reported.

The weekend's most startling news item, though, came Sunday
night on Walter Winchell's national newscast. Winchell reported that an
ex-paratrooper had entered a Luckie Street coffee shop shortly after the
fire and bragged of setting it because he believed his wife was in the
hotel with another man. Winchell's tip had come from Atlanta musician
Bill Cooper, so Solicitor General Andrews summoned Cooper to testify
before the grand jury the next day. Andrews said afterward that Cooper
had testified he overheard someone say the fire had been set. Andrews
dismissed Cooper's story, and there was never a follow-up article on
Winchell's report.

By this time L. P. "Pat" Patterson was becoming irritated at his
superiors at the Atlanta *Journal*, who, despite the flurry of rumors,
seemed as little interested in actively investigating the arson theory as did
city officials. He couldn't understand how his editors could be so
unconcerned about clearing up a major mystery.

To Patterson, the most obvious arson theory led to Robert
O'Connell, the hotel's burly operator. Patterson had learned that on the
day after the fire the Chicago *Tribune* reported that O'Connell, a policeman, had obtained a leave to look after the property of his brother,
Thomas O'Connell, who reportedly had operated an Atlanta hotel until
his death about a year earlier. The article also stated that O'Connell did
not live in Atlanta year-round; he had a Chicago residence at 6324
North Albany, where he lived with his wife and daughter.

Employees of the Winecoff wondered if O'Connell had finally irritated the wrong guy. Detectives asked Ward Duvall, the piano player in
the Plantation Room, if O'Connell had thrown anyone out of the bar
that night. Duvall said he couldn't remember because O'Connell threw
people out so often.

All the employees knew about the stocky former paratrooper who
had caused trouble in the bar the week before the fire — he could have

been the ex-paratrooper Walter Winchell referred to in his broadcast December 22. The young man liked to brag about how many Japanese soldiers he had killed in the South Pacific. The Winecoff's employees thought he was mentally unstable, perhaps as a result of a head injury suffered during the war. On Thursday, the night before the fire, he had entered the Plantation Room without a coat and tie, and the bartender had refused to serve him. He picked a big jar of boiled eggs in vinegar up off the bar and hurled it at the bartender, shattering a mirror. O'Connell was summoned and physically threw the man out on the street. But after O'Connell left, the man reentered the lobby and declared, "I'll get my revenge."

Patterson and George Goodwin, another reporter who had covered the fire, eventually wrote a memo to the *Journal's* city desk in hopes of persuading their editors to allow them to continue investigating the possibility of arson:

Materials, particularly the wainscoting and burlap cushion and draperies, have been tested to determine how flammable they were. All tests, including one made by state chemists on sisal fibers taken from the charred mattress, have shown no abnormal inflammability. . . .

Veteran hotel men and firemen say a burning mattress gives off a dense smoke and rarely bursts into open flame, yet the elevator operator detected no smoke on any but the last of her numerous trips to the upper floors. Neither did Mrs. Rowan on her trip to the fourth floor, nor the bellman and engineer on their trip to the fifth floor.

This does not explain why, within five minutes after the elevator operator discovered smoke, the lower floors were raging infernos of smoke and fire.

Patterson and Goodwin had investigated the Ku Klux Klan murders at Moore's Ford that year, and within six months Goodwin would win the Pulitzer Prize for revealing voter fraud in Telfair County. They thought the Winecoff fire could be a story of tremendous consequence.

Their article posed "some questions as yet unanswered." Among them were:

Why did O'Connell fail to tell Dunn & Bradstreet and attempt to conceal from a *Journal* reporter that he is a Chicago policeman on leave of absence?

Who is the bejewelled and well-dressed woman who frequently visited him here?

Why did O'Connell retire early on this particular night although it was his habit to constantly move about the hotel, supervising the work of employees, checking equipment and acting as house detective?

What truth is there in the rumor, circulated just after the Winecoff lease was purchased, that a Chicago gang had taken over the structure and would use it as a hideout and base for southern operations?

Patterson and Goodwin closed the memo with these words: "In the meantime, the investigations appear to be dying out, and it seems safe to predict that in a few more weeks the Winecoff Hotel fire will be forgotten."

CHAPTER 9

━ ≣◆≣ ━

T he investigations indeed died down within a month. It took longer than that for Rudolph Ogilvie to learn to whisper again.

Doctors had told him he'd never regain his voice, that his vocal cords had more or less melted together during the fire — probably during the few moments he was in the Winecoff's stairwell, stumbling and crawling from the eleventh floor to the twelfth. After six weeks, whispering was real progress. Ogilvie eventually was able to talk well enough to tell Lena Sollenberger what had happened. Her son Bobby, also a student at Milner High School in Barnesville, had been Rudolph's roommate that night. He told her how they had gone to the Henry Grady, where they had reservations but had been turned away and sent to the Winecoff. How Bobby had collapsed in Room 1126 and how Rudolph had been unable to find his body. How he was sure Bobby was dead before he ever left the room. But Ogilvie could never finish the story — about climbing a flight of stairs, opening a door, and falling over a woman's body — before Bobby's mother broke down and started crying. The loss of her only son was simply too great — so great that she herself lost her voice. Her husband took her to doctors around the country, but none of them could help her. After two long years — during which she tottered on the brink of a nervous breakdown — she was able to talk again. Then she started to recover.

Clara Strickland couldn't hold on long enough to get to that point. She was fifty-two years old and had been grieving for thirteen months, ever since her youngest daughter — sixteen-year-old Harriett, a Youth Assembly delegate — had died in Room 930. Early one January morning in 1948, in Georgia's extreme southwest corner, near the Spring Creek bridge halfway between Donalsonville and Bainbridge, her body was found hanging from a tree. Her car was parked nearby. She had driven from the family's 1,000-acre farm, five miles away. The coroner ruled it a suicide. The newspaper quoted the sheriff as saying she had been despondent since her daughter's death.

Ruth Baggett didn't make it through even a year. Her husband, Florence, the well-known livestock auctioneer, had died in 1112. Their children grown, forty-five-year-old Ruth was left to live alone in Claxton. Following back surgery earlier in 1946, she had become addicted to pain medication. When her doctor tried to get her off the medicine, she began to drink. Her three sons, unsure what to do, decided to seek treatment for her at Central State Hospital, Georgia's state mental institution in Milledgeville. Within a few weeks of her arrival, she was dead. Her sons were told it was heart failure. They felt the Winecoff fire had taken both their parents.

Ernest White — who had tried to rescue Winecoff guests like Florence Baggett — ended up in Milledgeville, too. Captain White had manned a net in the alley and been nearly knocked from a ladder by a falling body. He watched helplessly as other bodies crashed to the pavement. He then went inside and fought the fire all the way to the eleventh floor before being carried out to a fire department rescue truck by fellow firemen. By that time he had seen too much. His mind had snapped. The same phenomenon had afflicted the soldiers on the front line during the war; it was called battle fatigue.

Another fire captain, Furman Anderson, never recovered from the physical effects of the fire. A veteran of forty-two years with the fire department, he was sixty-three when he supervised Station 1's aerial truck crew on Peachtree Street. He directed the rescue of more than forty people, then went home and had a heart attack. He retired four months later and suffered through steadily declining health until his death twenty-six months after the fire.

The survivors were supposed to have been the lucky ones. But a never-ending replay of the mental pictures from the fire was a dubious reward. The images would return most often in the dark of night. Fireman Red Pittman didn't sleep for three weeks. When sleep would come, so would the nightmares. John Turk, who escaped from 1018, would sit up in the bed and start screaming. Catherine Weliver, who had been a woman of steel before the fire, scared of nothing, returned to Crawfordsville, Indiana, totally changed, unable to shake her memory of people running through flames on the fourth floor. Back home, she'd hear a siren and start quaking. It was a year before she could talk about the fire without crying. She never got another good night of sleep.

There were questions from well-meaning friends. Ed Kiker Williams was asked over and over about his mother, his sister, his aunt, and his three cousins, all of them dead. He preferred not to talk about the fire. Williams and many of the survivors were wondering themselves why they were still alive, why they had been lucky, why they had been spared.

And then there was the ballad. It was sung by well-known gospel singer Lee Roy Abernathy, who claimed he had planned to stay at the Winecoff that night but changed his mind and remained home in Canton, Georgia. Released eight months after the fire, "The Burning of the Winecoff Hotel" opens with the wail of sirens, then describes in three stanzas "the flames climbing upward," "the screams of agony," "the smoke and awful heat." Separating the stanzas is this refrain:

Those faces in the window, I see them in my dreams.
I always will remember those agonizing screams.
But surely there's a heaven for folks that die this way,
And we'll go home to meet them in glory some sweet day.

The song was considered in dreadfully poor taste but still made the airwaves, only furthering the fire's reputation as a disaster of almost mythical proportions, a latter-day Titanic. For those guests whose broken bodies lay in hospital beds for months, however, the fire remained very real, a living nightmare.

The Meyer sisters, Eva and Carrie, had sent a telegram the morning of the fire to their family back in Knoxville: "We are slightly injured and in Grady Hospital for treatment. Will not be home for Sunday."

Carrie didn't go home for six months.

The Meyers had jumped into a net from 902. Carrie broke her back, both legs, and both arms. Eva broke her back and arms.

Twenty-year-old secretary Daisy McCumber received even more serious injuries when she jumped for a net from the eleventh floor: compound fractures of both legs, a broken back, a broken pelvis, broken ribs, a fractured hand, a slight brain injury. She was not sure if she bounced out of the net or plunged through it.

While she was hospitalized, reporters identified her as the "jumping lady" caught in midflight in a picture that would win the 1947 Pulitzer Prize for amateur photographer Arnold Hardy. When shown the photo as she lay at Grady Hospital, she vehemently denied that she was the woman. But she later was identified as the "jumping lady."

McCumber eventually lost a leg to amputation and underwent one operation after another at hospitals around the country, all paid for the by Red Cross. At one point, she told a reporter, "I think I would rather die than go through all of it again."

Mildred Johnson, who fell into the alley when her sheets burned in two at the fifth floor, remained unconscious for two days. Her hand was paralyzed permanently.

Dorothy Moen, the senior from Baker Village High School, spent six months at Grady. Her mother caught the bus from Columbus every Sunday to come see her. Dorothy had knocked the screen from a window in 730 into the concrete court area outside Room 330, site of the poker game. When the smoke and heat forced her to jump a few minutes later, she shattered her leg and landed on the screen as she fell forward. The wire mesh worked its way out of her chin a year later.

Dorothy's roommate, Bess Dansby, recovered sufficiently from her fall from the window ledge of 730 to graduate second in her class and receive a four-year Georgia Vocational Rehabilitation Scholarship to Auburn University. Betty Huguley, who thought she was going to bunk with the Columbus girls but ended up staying by herself in 414, graduated as salutatorian at Spalding High School.

There was other good to be found in the fire's aftermath. The community of Cullowhee, North Carolina, built a house for Bill Bryson's widow, Vera, and their two little girls, Sherry and Sue. It was built on the farm where the Trailways bus driver had grown up.

In Thomaston a grassroots Bible study class sprang from a meeting of the families of the four high school girls who were found suffocated in 1430 with the room's Gideon Bible opened to John 14: "Let not your heart be troubled. . . . " The families started gathering a month after the fire, and by June the average attendance at the weekly meetings was thirty. As many as eighty people came to some meetings, and even Gov. M. E. Thompson attended one.

In Atlanta, 325 high school students from seventy-five schools attended the third Youth Assembly in 1947. It convened on the first Friday of December, just as it had in 1946, but instead of staying in hotels, the delegates slept on cots in classrooms at Georgia Evening College, later to become Georgia State University. High schools in Albany, Bainbridge, Columbus, Donalsonville, Gainesville, Rome, and Thomaston all sent delegates. The first motion on the floor was to ban smoking in the assembly; it was made by Charles Gray, who had been the last Rome boy out of Room 430 the year before, the one who glanced back to see the door fall in. Elected floor leader of the assembly in 1947, Gray lived in the same mill village in Rome as had Buzz Slatton, the 1946 clerk of the house who was killed in the fire.

On the final day of the 1947 assembly — Sunday, December 7 — in a 9:00 a.m. ceremony broadcast on radio from the third floor of the Capitol, a plaque was dedicated to the thirty students and two advisors killed in the fire.

The Atlanta Fire Department, meanwhile, became one of the best equipped in America, as Mayor Hartsfield essentially handed Chief

Styron a blank check after the fire. The Winecoff's absence of fire escapes and sprinklers became an impetus for building code revisions throughout the nation, with President Truman even calling a President's Conference on fire prevention in the spring of 1947.

Life went on. Dwight Lamar Morrison, Jr., was born two months and a day after his father died in Room 1026. Dwight Sr. had been looking for work back home, in the state he loved. He had piloted sixty-five bombing missions in a plane nicknamed "the Widow Maker" and then died looking for a job in the land he fought to protect. On December 7 the following ad ran under "Situations Wanted — Male" in the classified section of the Atlanta *Journal*:

Ambitious, energetic ex-AF major desires sales position with nat. known co.; college credits in business adv. Have car, free to travel; age 27. Call D. L. Morrison, Hotel Winecoff.

The oldest of five sons of a Methodist minister, Morrison had grown up all over south Georgia, moving every three or four years. He graduated from Middle Georgia College in 1940 and qualified for the Army's aviation cadets a year later. He flew a B-26 Marauder, a twin-engine plane with only a sixty-five foot wingspan but one of America's primary medium-range bombers. Joining up with the 9th Air Force in England in the summer of 1943, he bombed German installations in Holland, Belgium, and France. Usually flying lead for its group, Morrison's plane was shot up several times, but not a member of his crew was ever wounded, not even when five Nazi fighters attacked his plane over Rouen, France.

Morrison wore a little white cap for good luck and kept flying — long after he had logged enough missions to go home. By the end of the war, he had won the Distinguished Flying Cross and the Air Medal with nine oak-leaf clusters and become an operations officer, plotting bombing runs for other pilots. He came home to Tyndall Field in Panama

City, Florida, met a secretary named Hilda McMillian at Wainwright Yard there, and married her in March 1946.

Like Jimmy Cahill, Morrison was on terminal leave when he came to Atlanta Monday, December 2. He spent Friday evening with Nick Simmons, a high school buddy he had not seen since before the war. He planned to return to his parents' home in Lumpkin, Georgia, Saturday. Hilda, seven months pregnant, was waiting for him there.

William Todd's daughter was born six weeks after he died. He was a pilot, too, but unlike Morrison, he had spent the war years in the states, helping to develop America's fast-growing system of airports. An engineer with the Civil Aeronautics Administration, he was living in Atlanta when his father died suddenly in May. He had received a transfer to Columbia, South Carolina, so that he could look after his mother, but had returned to Atlanta to attend a meeting at the CAA's Region 2 headquarters. He had also purchased a surplus BT-13 trainer plane from the Army, which he planned to fly back to Columbia.

A parachute was found in a fireproof bag in Todd's room, 1214. It had burned, just like everything else. His body was identified by his 1927 Citadel class ring.

Todd had recently bought an additional life insurance policy but hadn't gotten around to signing it. His widow, Lib, and their three children moved into a low-rent housing project and got by on workmen's compensation.

Lib Todd and Hilda Morrison filed two of the more than 150 suits against the Winecoff Hotel Company. The suits came from not only victims' families but also injured survivors. The first one was filed exactly a week after the fire. The plaintiff was O. B. Torbert, father of Virginia Torbert, one of the four Thomaston girls whose Youth Assembly trip had ended in death in Room 1430. Because some sort of consolidation of cases was obviously needed, the first verdict would be critical.

The Torbert case finally came to trial in September of 1948, before Judge Bond Almand in Fulton County Superior Court, two blocks from the Capitol. The Torberts were seeking $15,000 from the owner and operators of the hotel — or their insurers — but the defendants' attor-

The Winecoff's burned-out hulk had remained a constant reminder of the fire during the twenty months that elapsed before the first civil suit came to trial in 1948. (The Mortgage Guarantee Building can be seen on the right.)

neys had scored a victory before the trial ever started. They had hoped to put two years between the fire and the start of a jury trial in order to allow public outrage to cool, and almost two years had indeed elapsed.

The Winecoff could hardly be forgotten, however. Its blackened, empty hull remained painfully conspicuous on Peachtree Street, where its sidewalk was still barricaded. Only its tall vertical sign had been removed — at the request of local hotel operators, who considered it a bad reminder for potential guests. Nevertheless, media interest in the nine-day trial was minimal. The Atlanta newspapers did not report daily on the testimony, and the stories that did run were usually buried inside the paper.

The Winecoff's owner, Annie Lee Irwin, was defended by Bill Grant. Hamilton Lokey was engaged by the Winecoff's insurer, Ocean Accident and Guarantee Corporation, to represent her as well.

E. D. Smith defended the operators, the Geeles and Robert O'Connell. They testified by deposition, having left the city shortly after the fire.

The Torberts' lawyer was Allen Post. His goal was to convince the jury that the Geeles and O'Connell were guilty of negligence, of running a "sloppy" hotel, and that Mrs. Irwin also was accountable. He argued that her ill-advised sale of the hotel to the hastily formed Arlington Corporation three days after the fire demonstrated that she was culpable.

Tom Branch, an associate of Post, pointed out to the jury that the Winecoff was advertised as "absolutely fireproof, yet had a saloon on the second floor, a gambling room on the third floor, with seventeen-year-old boys wearing turtleneck sweaters going back and forth between the two, and a fight in progress on the fifth floor, with a bellhop, who should have been watching for fires, carrying liquor to drunks."

The defendants' attorneys attempted to absolve their clients of any negligence by showing that the fire was arson. Their best evidence was the testimony of Asa Nuckolls.

Asa Nuckolls came to Atlanta — at Ocean Accident's request — on Saturday, December 21, two weeks after the fire. A consulting chemical engineer for Underwriters Laboratories in Chicago, he was the author of a book on combustion and generally regarded as America's foremost authority on arson. For four days — until Christmas Eve — he went over every square foot of the Winecoff. Then he ran tests on materials from the hotel and wrote his report.

Nuckolls' investigation and lab results convinced him that the fire was set by an arsonist, but his findings had been held for the trial and never released. Because of Nuckolls' heart condition, Lokey had sought to go to Chicago to take his deposition. But Ocean Accident believed Chicago lawyers could take the deposition just as well and far less expensively. It was a major tactical error. The deposition taken by the Chicago lawyers totaled 175 pages, far too lengthy to read to a jury. So the attorneys were able to present to the jury only Nuckolls' credentials and brief excerpts of his findings.

Nuckolls contended, for instance, that the flammable materials already in the hotel — the wainscoting, the carpet, and the mattress in the third-floor hall — "absolutely would not account for the effect produced by this fire." But despite their conclusive tone, the excerpts were simply not as convincing as a complete, well-constructed deposition would have been.

A frequent subject of testimony was the auxiliary fire hose stored accordion-like on the mezzanine and found lying unraveled after the fire. There was a two-inch slit along the crease near the end of the hose, along with thirty-two smaller cuts. Assistant fire chief Fred Bowen said the slit was caused by a guest turning on the water before taking the hose off the rack. Joe Mansell, an investigator for the solicitor, said he believed the hose, made in 1913, was rotten. Other testimony contended the slit occurred when the hose lay on the floor for twelve days during the cleanup of the building. Yet another witness said he believed the hose had been deliberately slashed with a knife by an arsonist.

The case for arson was further buttressed by the testimony of Warren Foster, the Columbus man staying in Room 508. He testified that his baby was awakened four times by fighting next door and that he heard a man say, as he left, "I'll get even with you before the night's over."

Investigators had been quite interested in Suite 510-512, especially since 510 was the most damaged room on the fifth floor. The two-room suite on the hotel's northeast corner was where Mobley, the bellhop, had delivered ice and ginger ale around 3:30, before beginning his inspection of the building with Fred Grennor, the night engineer. The room had been rented to Richard Fletcher and Harry Lindervall. Both were interviewed — Fletcher at Fort McPherson in south Atlanta, where he was payroll chief, and Lindervall at Fort Bragg, outside

Fayetteville, North Carolina, and again at Camp Stoneman near Pittsburg, California.

According to their story, Fletcher and Lindervall — a couple of Army buddies in their late twenties — happened to be on the same Delta flight into Atlanta on Friday, December 6. Fletcher was returning from a three-day pass to his hometown of Dayton, Ohio, and Lindervall was returning from a visit to his mother in Chicago. Lindervall told Fletcher he had wired ahead for a room at the Winecoff and invited Fletcher to join him for a drink. Fletcher, who had a fifth of whiskey with him, accepted Lindervall's offer; he didn't have to be back at Fort McPherson until six o'clock Saturday morning. They arrived at the Winecoff at about nine o'clock Friday night.

After having a couple of drinks in the room, the two men went down to the Plantation Room on the mezzanine about 9:45 and took seats at the bar. They each had about five drinks over the next couple of hours, then returned to their room with Dr. Joe Rankin, whom Fletcher described as a former Army officer and "friend of mine from Atlanta."

The three men continued to drink in 510 after room service had supplied them with ice and ginger ale. They talked about the war and about being overseas. About 1:30 Fletcher and Lindervall got into an argument. According to Fletcher, Lindervall made a disparaging remark about his wife, whom Lindervall had met in the spring while on furlough at Fletcher's home. The men's voices grew increasingly loud — loud enough to disturb little Connie Foster next door in 508.

Foster testified that the argument escalated into a fight, that it sounded to him like one of the men had the other on the floor and was "beating on him" when someone came up from the lobby to tell them to be quiet. And it was the fight, Foster presumed, that led to the threat issued by the man who left the room — Fletcher. In his original statement, Fletcher admitted that he and Lindervall fought, but three days later, in a longer statement, he said they had simply argued. Neither Fletcher nor Lindervall ever mentioned the threat.

Fletcher said that after he left the room he walked down Forsyth Street to the Ansley Hotel and caught a cab to Fort McPherson. He said he was back in camp by 2:50, three hours before his leave expired. "I was pretty noisy, and I think I woke most of the men in my room up," he admitted. "At reveille time, Harry [Lindervall] called me on the phone and told me the Winecoff had burned up. He asked me if it was possible for me to bring some clothes into him, and I told him I would try. He said he was at the Henry Grady Hotel."

By the time the trial came to court in 1948, Fletcher had been reassigned to Germany. Lokey was sure he had set the fire and wanted to go to Germany to interview him. But, again, Ocean Accident wouldn't pay the expense. The insurance company had only $80,000 of liability in the case and thought it might cost $10,000 to run Fletcher down. The company told Lokey that he could prove arson without naming an arsonist.

But arson was going to be especially difficult to prove, given the climate of public opinion. A jury that believed the fire was set by an arsonist would have freed the hotel's operators of responsibility and sent the victims' families home penniless at a time when they were still the object of much sympathy. From the morning the flames were extinguished, the attention of both the press and the public had centered more on the lack of sprinklers and fire escapes — an apparently gross omission for a "fireproof" hotel — than on the possibility of arson.

After less than two weeks of testimony, the jury got the case on a Friday afternoon, deliberated a while, broke for supper, and then reconvened at 7:00 p.m. A verdict was reached by 9:00. It was read by Post, the Torberts' attorney:

"We the jury, find in favor of the plaintiff and against the defendants, Geele, Geele and O'Connell, in the sum of $10,000. We find in favor of the defendant, Mrs. Annie Lee Irwin."

The jury had absolved the Atlanta resident — Annie Lee Irwin, the absentee owner who had sat through the entire trial — and blamed the fire on the Geeles and O'Connell, the outsiders. The verdict received only four paragraphs in the following day's *Journal*.

An agreement was struck between the lawyers for the plaintiffs and the lawyers for the four insurance companies that represented the Geeles and O'Connell. The remaining cases would be settled out of court. Fulton County judge Virlyn Moore would hear evidence as to damages on each case and then award an amount proportionate to that given the Torberts.

The awards totaled approximately $3.5 million. The operators, however, were covered for only $350,000, and Mrs. Irwin's $425,000 policy could not be touched. So each plaintiff received just ten percent of the court's award. The Torberts received only $1,000. Many plaintiffs got less.

CHAPTER 10

━━◄◆►━━

I nvestigating the scene during the days immediately after the fire, Ocean Accident engineer E. F. Otten had found the "trail of fire" that Asa Nuckolls would discover. Otten wrote:

It is assumed by the writer, and now generally agreed, that the fire was started by some person or persons unknown. . . . It is assumed that someone started at some point above the fourth, or at the fourth, floor and poured [an] inflammable compound or liquid on the floors around the elevator shaft and down the stairwell and around one side of the elevator to one side of the stairwell, where it was apparently ignited, the person escaping by descending to the lobby floor, where he could easily stay under cover and escape without being noticed.

Otten and Nuckolls were two of the four investigators who pointed to arson in written reports. The others were John J. McCarthy of the New York Hotel Association — who on December 15 had become the first to suggest arson publicly — and attorney Charles M. Huguley.

Huguley was a special agent for the National Board of Fire Underwriters, a highly respected investigator who specialized in determining if there was sufficient evidence to prosecute an arson case. He theorized before the grand jury, which convened on December 16, that the fire could have been "the product of the twisted mind of a veteran arsonist." He said the arsonist could have used the "streamer" method: the laying of a thin trail of gasoline from floor to floor.

Huguley, McCarthy, Otten, and Nuckolls had all seen the same evidence, so obvious that even a layman, *Constitution* editor Ralph McGill, had noticed it on his tour of the hotel hours after the fire had been extinguished. "At the third floor, about the elevators, there are signs of intense fire and heat," wrote McGill that day.

The wired glass in the third-floor door of the west elevator was melted into a glob — melted more than the elevator glass on floors where the fire was more widespread. A brass cuspidor near the elevator was melted to the floor. The carpet was completely burned. Even the concrete below showed heat damage. The area in front of the west elevator was the site of the fire's highest temperatures — at least 1,500 degrees Fahrenheit, the investigators believed. However, just a few feet away, in front of Room 314, the carpet was undamaged. Instead of spreading down the combustible carpet into the front hall, the fire had made a sharp turn up the uncarpeted steps between the elevators. The "trail of fire" wound around the east elevator on the fourth floor, then back up the stairs to the fifth floor, ending at another hot spot outside Suite 510-12.

The trail was visible because heat spalls concrete, leaving it chipped and flaked. The repair of this spalling effect — which would have occurred when the hotel was refurbished in 1950 — can still be seen today under the building's carpet.

"When you look for arson, you look for any abnormal spread because fire takes the path of least resistance. If it has equal opportunity to spread, it spreads equally," says current Fulton County arson investigator Bill Dodd, who has studied the trail out of professional curiosity. "The Winecoff's halls were H-shaped, so you should have equal damage in all corridors as the fire spreads upward. But on the third, fourth, and fifth floors, one side of the hall was burned and one side was not. Above the fifth floor, though, both sides of the hall were burned equally."

Nuckolls saw the trail immediately. A combustion scientist with the investigations of twenty-seven major fires or explosions to his credit, he was methodical and meticulous. He took samples of the debris in front of the west elevator on the third floor, put them in jars, and sent them to his laboratory at the University of Chicago. The vacuum-distillation tests there showed evidence of petroleum products, such as ordinary gasoline or gasoline jelly, a substance similar to the napalm used by the Army in World War II and Vietnam. Nuckolls also did spectrogram tests on the samples and found them to contain aluminum, magnesium, lead, and calcium.

"There must have been highly inflammable material in substantial amounts in the west corridor and particularly near the west elevator to start this fire," he wrote in his report.

The presence of foreign combustibles seemed to Nuckolls to be sufficient evidence of arson, but he knew a jury would need more proof. So he set out to determine if there were enough "heat units" available on the third floor — without the foreign combustibles — to account for the fire's swift spread. He simulated the conditions of the hallway exactly — right down to the Winecoff's "foamaway" rug cleaner, and the Red Arrow paste that adhered the burlap wainscoting to the plaster walls. He wanted to see if there was anything in the hallways that, when burned, could produce sufficient heat to melt the wired glass in the elevator door. He tested his recreated hallway in a "furnace tunnel," but he couldn't get the fire to become self-sustaining. When he took the torch away, the fire slowly fizzled out. This seemed to him to be indisputable scientific proof which no jury would be able to ignore.

But because Nuckolls' report was not made public until the trial twenty months later, it was a moot point. The public and, more important, Atlanta's government officials had already decided the fire was a terrible accident — even if basic laws of physics would have had to have been broken for that to be the case. Indeed, the sum of the investigations by Nuckolls, Otten, Huguley, and McCarthy could not equal the weight of a single report completed only two days after the fire. It was the initial report and the one that shaped the city's opinion.

The report was the collaborative effort of fire marshal Harry Phillips, H. N. Pye of the Southeastern Underwriters Association, and Carey Hutson of the National Board of Fire Underwriters. It was the report presented to the board of firemasters Monday, December 9, and, with the concurrence of fire chief Charles Styron, to the city council the following day. Phillips had put great credence in the accounts of the first firemen to reach the third floor — men like assistant chief Fred Bowen. To Captain Bowen, the inferno had all the markings of an accidental fire. He noticed the mattress left in the hallway and logically reasoned that, like countless other hotel fires, this one had begun in a mattress. He further assumed that a hotel employee had tried to fight the flames with the fire hose that lay unraveled on the mezzanine. As Bowen knew, this dangerous scenario was played out hundreds of times a year by hotel employees trying to extinguish flames themselves rather than immediately calling the fire department and thus having the hotel stigmatized as a firetrap.

So with the firemasters meeting Monday at 5:00 p.m. and Mayor Hartsfield — an ex-officio member — expecting an explanation of the cause of the fire, Phillips needed only to make a quick appraisal of the evidence that had been gathered Saturday and Sunday. There were the partially burned mattress, the mezzanine hose, and the report of Bowen and the other firemen. It was decided that the firemasters would be told that preliminary findings indicated the fire was of accidental origin. That would be welcome news to Hartsfield, who knew he would need public backing to pass more stringent fire ordinances and safety procedures, which he felt the city needed. Hartsfield felt an opinion of arson, on the other hand, could result in a sensational witch hunt, with no good to come from it. The press was appropriately tipped off.

The banner headline in Monday's Atlanta *Journal* read: "Live Cigarette, Mattress Blamed In Winecoff Fire." The Associated Press quoted Phillips as saying the fire had "probably been caused by an unidentified smoker who was dazed by liquor" and that the exact origin probably would never be determined for certain.

Solicitor General Shorty Andrews (right) and Joe Mansell, one of his investigators, examine the mezzanine hose during the 1948 civil trial. Mansell believed the hose had rotted, while other investigators felt it had burst under pressure or been cut by an arsonist.

While this conclusion sat comfortably with official Atlanta, it was clearly disputed by most other investigators. Eight of the twelve subsequent reports either doubted or completely dismissed the mattress theory, saying at the least that the fire was "of unknown origin." The National Bureau of Standards — which completed the longest and most thorough report other than Nuckolls' — even determined through testing that the mattress would have had to smolder for thirty-eight minutes before bursting into flames.

The Bureau of Standards suggested the presence of "other combustibles" but stopped short of calling the fire arson. Its thirty-page report also shot down the "delayed call-in" theory. Both the National Fire Protection Association and the Building Officials Conference of

America had suggested that a delay in calling in the alarm might have rendered the fire uncontrollable.

It's unlikely that there was any delay in reporting the fire, however, at least according to the sworn statements of night maid Alice Edmonds, elevator operator Rozena Neal, and telephone operator Catherine Rowan. None of them reported smelling smoke until the third floor was in flames.

The fire department reported receiving the Winecoff alarm at 3:42, which was consistent with the employees' accounts. The Bureau of Standards concluded that the fire reached an active flaming stage "with a considerable volume of smoke at 3:35 to 3:40, perhaps nearer the latter time than the former. There was apparently little or no delay in the transmission of the alarm to the fire department. . . ."

The burlap wainscoting on the wall behind the mattress in the hall was not even completely consumed, nor were the chair rail, the base-board, or even the mattress itself. Yet the flames supposedly spawned by the mattress reached 1500 degrees a few feet away.

Statements by employees and guests also pointed to arson. Alice Edmonds said the smoke had a funny smell to it. The Casey brothers — who let the trapped Edmonds into Room 304 — said the smoke looked whitish. Also, the first firemen to arrive on Peachtree Street said the third, fourth, and fifth floors were already in flames, further evidence of "foreign combustibles."

But fire investigation, born in the 1930s, was still in its infancy. Fire Marshal Phillips's fire prevention bureau — charged with enforcing Atlanta's fire codes, making inspections, and carrying out investigations — was simply overmatched by the enormity of its task. And two other local government investigations — by the grand jury and the police — would pay little heed to the untrusted new science and the evidence that the fire had been set.

In the same Monday, December 9, article in which Phillips advanced his careless-smoking theory, Solicitor General Andrews promised a grand jury investigation. The forty-nine-year-old Andrews, elected two years earlier after a decade as assistant solicitor, said he had met with both Phillips and Fire Chief Styron and would prosecute any-

Winecoff engineer Dewey Reed examines the roll-away bed and mattress left in the back hall of the third floor. Although the fire supposedly started in the mattress, it was not even totally consumed.

183

one proven criminally negligent.

Testimony from a list of more than a hundred witnesses began Monday, December 16. Andrews told the press the investigation would likely take two courses: one to learn whether city firefighting equipment and hotel precautions were adequate, and the other to determine whether the Winecoff's owners and operators had been criminally negligent in failing to provide fire escapes and fire-resistant stairway doors.

Andrews was quite interested in the tale of Rose Harvey. The University of Alabama senior had fled Room 1228 and pounded on the elevators before following Ralph Moody into Charlie and Millie Boschung's room — 1208 — from where she escaped on a sheet rope.

"The fact that Miss Harvey was able to run into the hall and remain for some time refutes the claims of the so-called experts that the hallways were death traps," said Andrews. "The additional fact that she made her own escape down a sheet rope proves beyond question that other guests could have saved themselves had a fire escape been installed along the side of the building."

Andrews heard Warren Foster testify about the fight in Suite 510-12 and the ensuing threat, but three days into testimony, he still told the *Journal* that nothing in his investigation up to that point "will support the theory of arson." He said "all evidence" on the fire's rapid spread pointed toward the "inflammable burlap and wainscoting, plus the terrific suction created by the open stairways."

During the week the grand jury had been in session, only three days had been devoted to Winecoff testimony. But with Christmas Eve a day away, the grand jury was ready to give Andrews his indictments. It was Monday, December 23, and the irony of the day was played out in the banner headline of the *Journal.* In the early edition it read, "New Evidence Presented Jury Supports Hotel Arson Theory." By the final edition it was, 'Fire Trap' Charges To 3 Hotel Lessees."

The indictments — one a felony charge of involuntary manslaughter and the other two misdemeanor counts of the same charge — were based on a 1910 city law requiring hotels with three or more floors to provide fire escapes and on the hotel's failure to have enclosed stairways. The article said the felony count could result in one to three years in prison for the three operators — Art Geele, Jr. and Sr., and Robert O'Connell.

After Christmas the grand jury spent its time arguing whether the fire department had been ill-equipped and poorly trained. Chief Styron commented that "In the case of the Winecoff fire, we were licked before we left the fire station." Nevertheless, the grand jury's presentments on January 2 criticized the fire department's tactics in fighting the fire and called for modernization of equipment and training, handing Hartsfield the official impetus to make his fire department exemplary.

Andrews had also gotten what he needed — a tidy case. At a time of political turbulence in Georgia, it was not expedient for him to spend any more time on the Winecoff.

Andrews was a crony of Eugene Talmadge, who had been elected governor in November after defeating Governor Arnall in the Democratic primary. Talmadge had planned to make Andrews a superior court judge, but the governor-elect died suddenly on December 21, plunging the state into what would be called "the year of the three governors." Andrews would still be in line to become a judge if Talmadge's son, Herman, became governor. And if the new governor were M. E. Thompson — the recently elected lieutenant governor who had spoken to the Youth Assembly — Andrews still expected to be a judge, because Thompson planned to promote his buddy Paul Webb, one of Andrews' assistants, to solicitor. For Thompson to do that, he would have to elevate Andrews to judge. So Andrews was in good shape either way — as long as he wasn't conducting a long, drawn-out arson investigation.

Andrews became a Fulton County superior court judge in April of 1947, appointed by Governor Thompson, who made Webb the solicitor. The indictments against Geeles and O'Connell were dismissed in 1948 by the Georgia Supreme Court, which ruled there was insufficient evidence to show criminal negligence. "Shorty Andrews was flamboyant and thrived on publicity," said former *Journal* reporter Aubrey Morris. "I thought the whole investigation was botched. There was a lot of publicity about the investigation, but no one went to jail."

In a 1985 interview, when asked how he thought the fire started, the eighty-seven-year-old Andrews replied, "My idea at the time was that it was some heavy drinking and a little fidoodlin' on the part of men and women, and it was purely accidental. I never got the idea that it was deliberately set by any one person. I know I thought it [the hotel] did not have enough means to get out in case of fire." Then in 1986 Andrews was interviewed for an Atlanta *Journal-Constitution* article reexamining the arson theories at the fortieth anniversary of the fire. "I investigated that fire thoroughly, and I could find no evidence that it was incendiary. If I could have found any evidence of arson, he [the accused arsonist] would have been indicted," said Andrews, who died of heart failure fifteen days after the article was published.

E. D. Smith, who defended the Geeles and O'Connell in the 1948 civil trial, believed the entire city government was extremely hostile to his clients. In a 1985 interview, he said, "I tried to get help from Hartsfield with the grand jury, and I couldn't get anything out of him. I think the city didn't like having a fire in downtown Atlanta that killed 119 people, and they were looking for a scapegoat wherever they could find one."

Hamilton Lokey, who defended Winecoff owner Annie Lee Irwin, agreed. "Somebody set that building on fire. I'm convinced of that," said Lokey, "but we got precious little sympathy out of the authorities and no sympathy from the newspapers."

On December 15 — the day it published McCarthy's initial statements theorizing arson — the *Journal* ran an editorial ("Let Us See They Did Not Die In Vain") that said evidence showed the fire was caused by "drunken revelers tossing lighted cigarettes carelessly."

By contrast, the *Journal-Constitution's* fortieth anniversary article quoted five Atlanta newspaper reporters as saying they had suspected arson but were "either distracted by other stories or discouraged by their editors from pursuing their own investigations."

But if the Atlanta newspapers and the Fulton County grand jury seemed to have little interest in probing the arson theory, the Atlanta Police Department may have had even less. Its official report discounted all theories of arson, blaming the Geeles and O'Connell for two acts of "negligence" — storing furniture in the third-floor hall and delaying in calling the fire department.

The report's findings were revealed in the same edition of the *Journal* in which Fire Chief Styron replied to a reporter's question about the fire's lightning-like spread. "I'm just as puzzled about that now as I was at the time of the fire," Styron said. "It just doesn't seem possible that so much fire could have resulted from one smoldering mattress."

The police investigation was conducted by detectives Jimmy Helms and George Slate, who worked closely with Fire Marshal Phillips, and written by Slate. "Slate really wasn't much of an investigator," said former police chief Herbert Jenkins in a 1988 interview. "It was obvious to us that they weren't going to find an answer, and all we wanted to a certain degree was — like Shorty Andrews — to get as much as you can on paper that looks good and get it out of the way. That was really what the police department wanted out of it."

Jenkins, who was a watch commander at the time of the fire, became police chief February 2, 1947, succeeding Marion Hornsby, who had died of a heart attack earlier that day. One of his first actions as chief was to challenge the power of the Ku Klux Klan-controlled police union. He later kept the peace during the turbulent 1960s — and helped Atlanta earn a national reputation for racial harmony — by defying state politicians who were writing massive-resistance segregation laws only a block from City Hall. He was the lone southerner named by President Lyndon Johnson to the National Advisory Commission on Civil Disorders in 1967.

Slate is dead, and his report on the Winecoff no longer exists. The police department did not keep criminal investigation files then. The report would be in the grand jury's file if that file had not been destroyed in a courthouse storage room fire in the 1960s. So of the three official investigations — fire department, grand jury, and police — only the one prepared by Fire Chief Phillips still exists.

The work of Ocean Accident's persistent Frank Hollman is therefore the only surviving investigation into who might have set the fire, and it exists only in part. A former Army intelligence officer, Hollman was claims superintendent of Ocean Accident's Atlanta office, and he worked doggedly on the case, as did Lee Hunter, the assistant claims superintendent. But their official reports have been destroyed as part of the company's routine file destruction process. Hollman is dead, and all that remain are copies of his individual reports in attorney Hamilton Lokey's case files, which have been given to the Atlanta History Center.

Interestingly, none of Hollman's reports delves into the big poker game in Room 330. "I don't believe I ever got into any evidence on the poker game," said Hunter in 1993. "We never could pin down any motive for someone in that game to start the fire. I don't believe we ever knew who those people were. If we'd had any leads, we'd have pursued them."

To do so, however, could have hurt their client in court. How culpable would the Winecoff's owner, Annie Lee Irwin, have been if she had leased the hotel to operators who rented rooms to professional gamblers and let them run all-night poker games?

——— ᴇ♦Ӟ ———

There was nothing unusual about a big poker game taking place in a room of one of Atlanta's better hotels. Gambling was a hotel staple, and not just in Atlanta. In other southern cities — in New Orleans and Montgomery, for example — floating crap or poker games operated around the clock in hotels, always on a lower floor so there could be several ways to escape.

In any hotel in Atlanta a professional gambler could rent a room and then invite in several cronies, but investigators believed that the game in 330 was the city's floating game, the one that moved from one hotel to another every few days, always staying just ahead of police. At the Winecoff, 330 was the ideal site since it was the biggest room on the lowest floor with guest rooms.

So the city had legitimate reasons for not pursuing any arson theories related to the poker game. If 119 people had died as a result of a floating gambling game the city was well aware of, political turmoil could have surely followed.

As recently as late September, Shorty Andrews and two other Fulton County officials had declared publicly that organized gambling was at its "lowest ebb" in recent years. Their remarks were reported in the *Journal* apparently to reassure the citizenry after two well-publicized gambling-related incidents. C. A. Wallace had been murdered during a marathon poker game at his bathhouse that month, and in August thirteen men had been arrested at Ponce de Leon Park during a Southern Association baseball game between the Atlanta Crackers and the Chattanooga Lookouts. Detectives had made the arrests in an effort to eliminate what was described as a "carnival-like atmosphere" in the left-field bleachers, where gamblers were giving four-to-one odds that the batter would not hit a flyball to the outfield.

"There's lots of gambling going on in Fulton County," Andrews said in the article. "But organized gambling, by professionals operating an entire house or hotel suite, is practically nonexistent."

When the poker game in Room 330 was first reported by the *Constitution* six days after the fire, detective superintendent E. I. Hildebrand said the police learned of the game from a clerk at another hotel — not from the Winecoff operators. The clerk said a professional gambler who lived at his hotel had left a message for his wife that he could be reached in a certain room at the Winecoff. That tip revealed the professional nature of the game, but it shouldn't have been the first clue that gambling was going on in Room 330. Both Alice Edmonds, the night maid, and Rozena Neal, the elevator operator, gave early statements about the cursing and shouting and coming and going in 330.

Edmonds — who was stationed in the linen closet next to 330 so she could receive assignments from the front desk — told authorities two days after the fire: "There was a crowd of men in 330, making a lot of noise, coming in and out of the hotel."

Edmonds also described the half-hour — from 2:00 to 2:30 — when she relieved Rozena Neal:

> While I was operating the elevator, I carried three men from Room 330 — I had seen them before — down to the lobby. One of them had on a blue overcoat. He was shorter than the other two men. One was tall and heavy, one had a scar on his face. . . . A little before Rozena relieved me on the elevator, I took the same three men from Room 330 back up to the third floor, also a young man

with a light beige overcoat and white pants. He was with the other three men. The short man seemed to be drinking. He asked me if I drank.

Rozena Neal later identified the "young man" as Harry Gilbert, a seventeen-year-old North Fulton High School cheerleader still wearing his white cheerleading pants from North Fulton's basketball game at Druid Hills earlier that evening.

Neal said that about thirty minutes before the fire she took two men — "one dressed in a brown suit, brown hat, light topcoat" — up in the elevator. "I don't remember the floor," she said. "He asked me if the place caught on fire, how would anyone get out. I told him there was the steps behind the elevator. He asked me where were the exits. The steps was all that I knowed of. He looked around the No. 1 elevator, as if he was looking for the steps, when I shut the elevator to go back down."

Neal also told of hearing loud talking in 330 while she was in the third-floor linen closet during her "lunch break" between 2:00 and 2:30 that morning.

There were also the statements of the four boys from Rome in 430: they heard shouts and curses from 330 for some time before the fire, and when they escaped to the roof of the adjacent Chandler Shoe Store, the poker players were there, too. The men had pushed the boys back so they could be the first ones down the fire escape at the back of the roof.

The December 13 *Journal* quoted Detective Helms as saying he had questioned three men who had been in 330, but that they all said they had only been drinking, not gambling. The detective also said he had the names of eight professional gamblers who supposedly had been in the hotel that night. The following day the *Constitution* reported that Slate was holding playing cards and whiskey bottles found in the room but that detectives were having difficulty rounding up the participants. Helms said he had been told two of them had gone to Miami. Superintendent Hildebrand was reported to have ordered an investigation to determine if the poker players could be prosecuted under gaming laws. On Sunday, December 15, the *Journal* said eight poker players had been questioned.

"We picked up everyone involved in the 330 game, and they all denied having anything to do with the fire," said Helms in a 1985 interview. "It was hard to find those gamblers."

On December 16, the day the grand jury began hearing testimony on the fire, five professional gamblers were scheduled to testify. If they did, it was never reported, and with both police and grand jury files lost, there's no way to know for certain who, if anyone, was questioned. Sketchy memories are the only remaining guideposts.

"There were allegations about a lot of shady characters in that room that night," said Aubrey Morris, the *Journal's* police reporter at the time.

But Detective Helms could recall only two names and a nickname. "I think one of them was called 'Go Lightly,'" he said. "And I know King Luckey and his brother were there."

Frank and King Luckey had been arrested in May of 1946 at a Mitchell Street poolroom after they were identified as robbing a Lawrenceville filling station of $300 two days earlier. Arrested with them at the poolroom — although later released — was Eugene Lamar Black, thirty-four, of Capitol Avenue.

Lamar Black was one of the two men who had rented Room 330 on Wednesday and were still occupying it Friday night. The other name on the hotel registry was Frank White. The two men listed their hometown as Covington, Kentucky.

Although none of the gamblers' names surfaced in the newspapers, their past deeds did. The second paragraph in the initial December 13 *Constitution* article stated: "Several of the players, reports indicated, also were in the alleged poker game on September 8 at 1400 Peachtree St., NE, when C. A. Wallace, bathhouse operator, was shot to death."

The murder was committed during a Sunday afternoon poker game that had started the previous evening. After playing a few hands, one of the participants pulled a gun and held up everyone in the game. Wallace tried to stop the robbery and was shot to death. The gamblers claimed the gunman was an "outsider" and said they didn't even know his name. Three months later, no one had been charged with Wallace's murder, but Andrews had gotten indictments on ten people on gambling charges. One of them was City Councilman Joe Allen.

However, as soon as the *Journal* hit the street December 13, Detective Superintendent Hildebrand denied any of the players in 330 had been in the bathhouse game.

Investigators tried to learn more about the 330 game from Harry Gilbert, the cheerleader. Gilbert said he was at the Varsity, a drive-in restaurant near Georgia Tech, around midnight when he ran into Walter Leonard, a friend of his older brother's. He said Leonard invited him to the Winecoff to talk about "old times." Beyond that, authorities said they were unable to obtain a full statement about Gilbert's presence in Room 330.

In a 1985 interview, Gilbert refused to discuss who else was in the room that night, referring to them as "rough customers" and then saying, "I don't remember their names." He said he saw no friction in the room. "As far as I know, no one in that room had anything to do with that fire. There was no argument, and no one was coming and going," he said.

In 1986, in the fortieth anniversary article in the *Journal-Constitution*, Gilbert — interviewed on the condition of anonymity — said he "would rather it all to be forgotten." The story ended with a Gilbert quote: "I put it behind me. I figured those [investigators] were a lot smarter than I, and whatever cause they came up with, I assumed that was it."

But in interviews between 1985 and 1991, four people — a detective, a grand jury witness, and two reporters — said investigators suspected that one of the gamblers set the fire intentionally.

One of the four was Helms. "I never believed it was arson, but it could've been that one of the gamblers up there got mad with the other ones and started the fire. That was one theory," he said, "but there was no evidence to that effect. I believe even if it [the fire] was from the boys in that room, it was just careless. That was my opinion, that it was just an accident."

Paul Lankford, the grand jury witness, was questioned extensively by investigators since his room, 324, was so near the flames' origin and he had heard rattling at his door. "They were suspicious of the poker game, but they didn't know that much. They thought it was a floating poker game," Lankford said in 1991. "The fire marshal took me back through the hotel. He told me there were traces of petroleum products and magnesium in the debris. He said a magnesium bomb would not make any noise."

"It was the card game. Somebody got thrown out and then came back," said longtime *Constitution* police reporter Keeler McCartney in 1993. "But I never could nail it down, and neither could the police." McCartney was the first reporter to write about the game in 1946. He said the Winecoff was one of the two most frustrating cases he worked on during his more than three decades with the newspaper.

"I never got that poker game off my mind. It sort of burned me," he said. "You'd get a lead, but then you couldn't pinpoint where it came from. You'd find one of the gamblers, and he'd say, 'No, I didn't say that.' Somewhere there was a simple answer if we could have found it."

"Here's what you ran into," explained McCartney. "Those were professional gamblers, and what would run foremost in one of those guys' minds is he might not be sure what the cops would do to him, but he knew what the gamblers would do if he talked about them. They'd kill him. It just stands to reason that faced with those kinds of odds a guy's not going to do anything but shut up."

During a 1985 interview, Aubrey Morris, the former *Journal* police reporter, recalled a poker game-related scenario similar to the one McCartney described: "I remember the stories that allegedly someone in the poker game got mad and set fire to the place," he said in 1985.

⊷ ⧯ ⊶

That person may have been Roy McCullough. Atlantan Ray Maddox, a former convict, names McCullough as the arsonist in a brief one-paragraph passage in his self-published 1981 autobiography, *In This Corner*, coauthored by Atlanta clergyman Gene Moffatt. Maddox, a born-again Christian and a deacon at Atlanta's First Baptist Church, refers to McCullough — nicknamed "the Candy Kid" — as simply "Candy" in the passage:

Candy had burned the whole [prison] camp down, when we were in, just to get to one man. He did the same thing years later when he set the Winecoff Hotel on fire, because he remembered one of the men in a gambling game we had hit there as the one who had ratted on him in another prison experience. We tried to stop him, but he went back and set fire to the hotel.

McCullough, thirty-seven, had spent much of the previous two decades on chain gangs and in Georgia's prisons, having been first arrested in Atlanta at age seventeen. He had, in fact, been in the Fulton County jail that very week after a brazen midday robbery of the Pig 'n' Whistle Friday, November 29, exactly one week before the fire. McCullough and two accomplices had walked into the crowded Ponce de Leon Avenue barbecue drive-in at lunch hour and walked out with $3,717.

McCullough was arrested the next day after his photo was identified by witnesses. He was charged with suspicion of robbery, but because a suspect could be held for only seventy-two hours, he was back on the street again by Tuesday, December 4. Before he could be charged with robbery, the grand jury had to return an indictment against him — the same grand jury that would investigate the Winecoff fire two weeks later. The indictment wasn't returned until Friday, December 6. The accompanying warrant for his arrest was then turned over to Fulton County deputy sheriffs J. M. Turner and Charlie Maddox. Police records show that McCullough was arrested, booked into jail, fingerprinted, and photographed sometime Saturday, December 7, just as investigators were beginning to sift through the Winecoff's ashes. A two-paragraph article on the indictment of McCullough and his accomplices appeared in Sunday's *Constitution*, adjacent to the coverage of the fire.

The prison camp fire had been at Ben Hill in southwest Atlanta eighteen months earlier — on June 7, 1945. Damage was estimated at $15,000, but of the 125 prisoners in the camp, the only victim was Austin Watkins, serving a life term from Forsyth County. The Atlanta *Journal* reported that Watkins was sleeping near the mess hall and quoted prison officials as explaining that the fire had started when a cook —

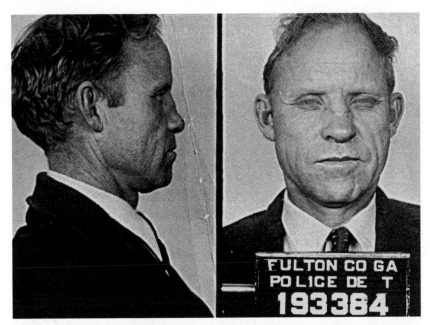

These mug shots of Roy McCullough were taken upon his arrest the day of the fire. He had been indicted the day before for robbing the Pig 'n' Whistle on November 29.

who was supposedly making gravy by stirring flour and pouring water into hot grease — saw flames in his frying pan and set it aside, catching the woodwork on fire. In his book, Maddox says McCullough started the fire because Watkins had "ratted" on someone who had been sent to the electric chair.

Maddox was twenty-two at the time of the Winecoff fire and had been out on parole from Georgia State Prison in Reidsville for four months, after serving two years for robbing an Atlanta gas station. He had grown up in Atlanta and had graduated from one type of crime to another, starting — by his own admission — with purse-snatching when he was ten. By the time he turned eighteen, he was in Georgia Sate Prison at Reidsville for auto theft. That's where he says he met McCullough. After McCullough was paroled in the summer of 1942 and Maddox escaped, Maddox says they became part of a loose group of criminals who would eventually call themselves "The Georgia Boys." Their game was robbing other criminals; they knew criminals couldn't call the cops.

Maddox says the robberies were directed by Leo B. Nahlik, the same man who had taught him the scam while Maddox was still a young teenager working at his father's popcorn stand at Ponce de Leon Park. Maddox says he and his friends knew who the "flyball" gamblers were and would rob them as they left the game.

Nahlik was an Atlanta policeman, considered by many to be the city's top detective. In 1942 he took a leave of absence to join the Marine Corps but remained in Atlanta as a Marine recruiter. Upon his discharge in 1944, he was named police chief in suburban DeKalb County. He abruptly resigned — allegedly under pressure — in February of 1946. One week after the Winecoff fire he rejoined the Atlanta police department as a detective. The December 15 Atlanta *Journal* article that reported Nahlik's return to the force said his partner would be George Slate. Five days later, Slate turned in his report on the fire, dismissing all possibility of arson.

Nahlik died in 1969. McCullough died in 1964.

Maddox offers no explanation in his book of how McCullough set the fire. Because of the difficulty of sneaking a container of flammable liquid through the lobby, current Fulton County arson investigator Dodd assumes an arsonist would have had to use a combustible already in the building. The most obvious possibility would have been mineral spirits, which were used heavily in painting at that time and were in abundant supply in the basement, according to 1948 trial testimony. Mineral spirits also could have been temporarily stored on a hall that was being painted. Several guests, including at least one on the third floor, told investigators that they had seen workmen painting Friday or earlier in the week.

The broken glass of a hotel water pitcher was found in the debris around the third-floor elevators. According to attorney Hamilton Lokey, arson investigator Asa Nuckolls believed the arsonist used the pitcher to spread his accelerant and then threw it against the wall. However, as Dodd pointed out, to leave a trail winding from the Winecoff's fifth floor to the third could have required as much as five gallons of a kerosene-based accelerant like mineral spirits.

An arsonist would have had to leave the building through the lobby, going right past night clerk Comer Rowan. But if Rowan saw anyone that aroused his suspicion, neither he nor his wife told investigators, and both were called to testify before the grand jury. Well into middle-age then, the Rowans are believed dead now.

Harry Gilbert told investigators that while he stood on the roof of the Chandler Shoe Store building, he saw a suspicious-looking man walking down Pryor Street, past the Loew's Grand and Bell Isle Garage. Gilbert thought it was peculiar that — despite his loud shouts of "Fire!" — the man just kept walking, turning left on Houston Street and disappearing.

Gilbert also told investigators he saw a shirtless drunk man staggering in the back hall of the third floor as he was leaving Room 330 around 3:00 a.m. He said when he returned minutes later — to retrieve his watch, which he claimed he had forgotten — the man was staggering up the steps from the mezzanine.

Earlier, around 2:30, Plantation Room cashier Alpha McCrary said she was making her nightly report at the front desk when she heard someone hurrying down the Winecoff's marble steps. She told investigators:

> A man came down the stairway and went out the front door. He kept looking back as though he thought someone was following him. This man wore a gray suit and gray felt hat. He was about five-feet-six, weighing about 140 to 150 pounds, no overcoat, hair gray at the edges, long face, nothing in his hands. . . . What attracted my attention was his hurrying and his looking back. I had never seen this man before.

Maddox said in 1992 that he doesn't know who McCullough's target was, although he claimed in the book that the man was one of the fire's 119 victims.

Given the path of the "trail of fire," bellhop Bill Mobley seems the most likely target — for three reasons. The most obvious is that the "trail of fire" led to Suite 510-512, where Mobley was drinking with Sgt. Harry Lindervall. Second, Mobley had been in Room 330 that night. And third, Maddox wrote that the target had "ratted" on

McCullough in prison, and Mobley and McCullough may have served time together in 1944, either in jail or on a chain gang. In June of that year Mobley was given a six-month sentence for assault and battery. McCullough was sent to the chain gang five months later after receiving an eighteen-month sentence for stealing a car.

"My old man liked to fight," says Bill Mobley, Jr., whose father died in 1979. "He'd fight anybody anytime. I saw him get into a couple fights that I'd run from. He'd intimidate people and try to pick fights."

Bill Mobley, Sr., was thirty-two years old and living in West End at the time of the fire. Investigators learned he was sent to 330 that night to quiet things down, just as bellhop Wayne Wise had done before Mobley relieved him at eleven o'clock. "Daddy said they had a card game up there and it got out of hand, and a couple of guys had pistols," said Bill Jr.

Bob Tweedell, the Winecoff's daytime desk clerk, suspects Mobley would have spent some time in 330 anyway. "If there was a poker game going on, he'd have been involved in it. He was a big gambler. He'd gamble on anything — football games, whatever there was to bet on. He'd do anything to make a buck," said Tweedell.

"My father loved the hotel business," said Mobley. "He got out of it a couple of times but then got back in it and stayed in it all his life. He knew how to make money in it and what to do and what not to do."

Mobley said his father believed the fire had been set because of the way it "rolled" up the hall at him when he opened the door in 510. Having been in both 510-512 and 330 that night, Mobley thought the fire's source was the poker game. "He never talked about the soldiers starting the fire; he always thought it was the poker game," the younger Mobley said. "He said it had to have been set, that the best he could tell, it started on three floors at the same time. He said somebody would have had to use gasoline or something for it to start so fast. He thought somebody had run down the hall with gasoline."

Mobley said his father was tormented by the fire the rest of his life. "When I was a kid, I'd wake up in the middle of the night and hear him screaming, 'Fire,' in a dream. He had nightmares until he died. I think he felt responsible because he was the bell captain. These people were his charge. He couldn't get over that there was nothing he could do," he said. "I think it made him drink a lot. My mother said he drank about twice as much after the fire. He couldn't go to bed until he had six or eight beers."

Mobley, still working as a bellhop, was arrested in 1965 for selling whiskey and in 1968 for selling whiskey and aiding and abetting prostitution. His son said he died an alcoholic at age sixty-four.

If Mobley was not the target, however, someone in the poker game might have been, and McCullough could have started the trail of fire on the fifth floor simply to spread the flames over a greater area and ensure that they would not be easily quelled.

Another possibility is that the target was Robert O'Connell, the burly Chicago cop who was one of the Winecoff's operators and its house detective. That theory would have to be based on McCullough's supposed relationship with Nahlik and on the words of O'Connell's associates at the Winecoff.

Nell McDuffie, the hotel's assistant manager and executive secretary, said in a 1987 interview that she heard before the fire that O'Connell had made numerous enemies among Chicago gangsters and had received threats since coming to Atlanta. She said Arthur Geele, Sr., one of O'Connell's operating partners, had told her: "I think Robert is going to have to be real careful because somebody's after him."

Geele's wife, Esther, reaffirmed the family's suspicion of O'Connell when she wrote McDuffie after the fire. In one paragraph of the letter, she said:

> My daily prayer is 'Who sabotaged the Winecoff? Set it on fire for revenge.' This is and always will be my opinion, as many threats had been made against Mr. O. due to the fact of expose of stealing going on by colored help in the Winecoff and through these investigations exposed bug racket ring to lottery ring. And we the innocent victims of circumstances. God answers prayers.

Nahlik was known by the police and criminals as the "banker" for one of Atlanta's postwar illegal lottery rings. If O'Connell had indeed helped bust a ring, Nahlik could have conceivably paid McCullough or one of his other stooges to knock him off or, to send a louder message, to destroy his hotel.

"I always thought Leo could and would — and did — do anything," said former police chief Jenkins, who died in 1990. "I thought he was as crooked as anyone on the police force, but I couldn't prove that. He was a smooth operator. He was pretty much known as a gambler. There's no question that he got into the bug [lottery] business after I became chief. I got word several times that he had a group of people writing lotteries for him. Hildebrand [chief of detectives] tried his best to catch him. He would raid a place, but Leo would always get out just before he got there."

Walter Leonard — the man who supposedly ran into Harry Gilbert at the Varsity — believes McCullough was in 330 that night. "I suspect he was there," he said in a 1991 interview. Asked if he knew McCullough, Leonard replied, "He was a hustler. I know he was one of those guys there."

Leonard was thirty-two years old then, a resident of Capitol Homes and a partner in a small used-bag business. He said he and his business partner, Burl Smith, heard about the poker game that night from the Luckey brothers and "a guy named Black" at Davis Brothers Restaurant, an all-night eatery on Luckie Street. He said he knew the Luckeys from York Pool Room at Pryor and Houston.

Leonard described the men in 330 that night as hustlers. "They hung around uptown there — at York's Pool Room and down on Luckie Street. They never worked, just hung around uptown and tried to beat everybody out of any of the money they'd get. They gambled all the time," he said. "They was bookies, too. They used to book flyballs out at the ballpark when the Crackers played out there."

Leonard said as many as a dozen people were in 330, that the game was five-card stud, with six men playing at a time around a big table. Everybody was drinking. He said he has no recollection of the game being robbed. Told of Maddox's account of how the fire started, he replied, "I'm not saying what you say isn't true, but there wasn't any robbery in that game."

Maddox — a retired security guard with the Georgia Building Authority and chairman of the offering committee at First Baptist — is evasive in discussing whether McCullough robbed the game or played in it. In 1986 he claimed he had not personally participated in the robbery and was unsure if the robbery had occurred that night, the night before, or in November. He also said McCullough gambled in 330 regularly. But in 1992, Maddox said he did try to stop McCullough from returning to the Winecoff when the two of them were at Thompson's, an all-night cafeteria on Peachtree about four blocks from the hotel.

Harry Gilbert still refuses to discuss the poker game, possibly out of fear born in the days following the fire. On January 8, 1947, in a signed statement, he told insurance investigator Frank Hollman: "When I was down in the grand jury, while out in the hall, one of the men in Room 330 asked me what I was going to tell. I told him I was going to tell the truth. He said there was no need to say we were gambling because we were not. They were not gambling while I was there."

When Gilbert was told Maddox's account in 1992, he said, "Let's leave it be. There's no use slapping this all over the papers. If it came up in a lawsuit, that's a different story, but I'd have to even think about that. I'm going let sleeping dogs lie."

Maddox's story about the robbery of the game is confirmed by Joe Mansell, an investigator in the office of solicitor general Shorty Andrews. He said he interviewed a man from Dahlonega who said the game had been robbed. "This was an older guy, and he said three guys came into the room, all with guns, and told them to unload their pockets. They made them stand in the corner," Mansell said in a 1987 interview. He said he didn't remember the man's name.

The strongest confirmation of Maddox's story, though, comes from Mansell's recollection of three young men he saw lingering around the Winecoff during the days following the fire:

"I asked them, 'What y'all doing up here?' And they said, 'Oh, we're not doing anything. Just looking around.' I said, 'They've got it roped off out there. Did you have to get permission?' They said they didn't need it. One of them said, 'We don't get permission where we go.'"

Mansell didn't like any of them — he thought they were punks — but he decided to keep listening. He was in plainclothes so they didn't realize he worked for the solicitor general.

"The one doing all the talking said, 'You've got a dumb bunch of policemen in this town. They ought to know that this didn't just happen. They can't figure out who did this,' " said Mansell. "So I asked him, 'Could you figure it out if you were a policemen?' He said, 'I could have already told you who done it.' "

Mansell tried to get the man to admit he had set the fire, but the other two would tell their friend to shut up when they feared he was saying too much. Mansell told his boss, assistant solicitor Paul Webb, that he thought the man had set the fire. They identified him from police photos and found he had a long criminal record in Atlanta.

"Paul asked me, 'How you going to prove it? Arson is hard to prove.' I told him, 'We need to put the squeeze on this guy and make him talk.' He said, 'You can't make a guy like that talk. This guy is smarter than we are because, as far as crime goes, he knows what to do and what not to do.' I said, 'If you gave him enough money, he'd squeal. Or else one of the other two would,' " Mansell said.

But Webb was not interested in bribing the man or his friends; he was afraid such a move could destroy his case in court. He had been charged with prosecuting the case against the Winecoff's operators, and within months he would succeed Andrews as solicitor.

Mansell does not remember any of the men's names and said all his notes were destroyed. "They cleaned out the evidence room every seven years. I tried to take the files with me, but they wouldn't let me," he said.

Mansell does recall, however, where one of the men said they lived: Formwalt Street, which was in a working-class neighborhood south of downtown. Maddox's uncle and his supposed mentor in crime, Bob Maddox, is listed in the 1946 Atlanta directory as living at 752 Formwalt Street. Maddox, at twenty-two, could have been living with his uncle or simply using his address.

Mansell could not remember if his conversation with the three men occurred the morning of the fire or two weeks later, while he was assigned to assist arson investigator Asa Nuckolls during his four-day inspection of the Winecoff. If the conversation had been December 7, McCullough could have been among the three early in the day. Two weeks later, McCullough could have conceivably been out on bond; Joe Lee Bishop, one of McCullough's accomplices in the Pig 'n' Whistle robbery, was freed December 21 on $10,000 bond.

Seven weeks after the fire — January 24, 1947 — McCullough came to trial for a month-long robbery spree that culminated at the Pig 'n' Whistle. As was his style, he pleaded guilty. He received two ten-to-fifteen-year sentences and two fifteen-to-twenty-year sentences. He was turned over to state prison authorities February 5, 1947.

CHAPTER 11

R oy McCullough was born in 1909 in Fayette County, just south of Atlanta. He was one year old when his father, Andrew McCullough, was charged with assault with intent to murder after firing several rifle shots at a black woman. He initially fired into the woman's home, then continued the fusillade when she took refuge behind a huge wash pot in her backyard. He got off with a fine.

A year later McCullough's father murdered a black man with whom he had been gambling in an all-night "skin game" on W. J. Bridges' plantation in neighboring Spalding County. The game occurred in the woods in front of the McCulloughs' tenant house. When a dispute arose, Andrew McCullough pulled out a shotgun, killed Walter Reeves, and seriously wounded Mitch Redding, another black man. He was convicted of voluntary manslaughter, sentenced to twenty years in prison, and disappeared from his son's childhood forever. The next time they saw each other, both were in prison.

Mittie McCullough, Roy's mother, moved her six sons and two daughters into Griffin and went to work at the Georgia Cotton Mill. In

1921 she was killed by McCullough's seventeen-year-old brother, Raymond, at their home in the mill district. Raymond claimed the shooting was accidental, but family members believe it was intentional. Newspaper reports at the time indicated the two may have argued over $150 that Mittie was holding for an older son, Alvin. Raymond had the money after the shooting, but no one was charged in the incident.

McCullough, twelve at the time, went to live with various relatives but never stayed anywhere very long. He quit school in the seventh grade.

He became the Candy Kid in 1926 when he and four other youths broke into the Sophie Mae Candy Corporation just west of Georgia Tech and stole one hundred pounds of candy. Four weeks later McCullough was given a $500 fine and six months of probation, the terms of which were very specific. He was to work regularly, be at home by eight every night, not go to any movies, and not associate with any of his codefendants. For McCullough it was a demanding sentence; within five months, his probation was revoked. He was ordered to serve the remainder of the original twelve-month sentence on the Fulton County chain gang. He was still not eighteen years old.

McCullough returned to the chain gang nearly a year later for a more serious offense: stealing a Dodge. By the time he was twenty, he was in state prison, serving a four-year sentence for "assault with intent to murder." His vicious attack on an elderly man exhibited a meanness that the court was determined to curb before he was any older.

The attack occurred May 14, 1929, at the corner of Harris and Williams, two side streets in downtown Atlanta. C. H. McHan, an eighty-four-year-old coal salesman, was calling on a customer when he reprimanded McCullough for the language he was using in front of women. McCullough threw McHan to the sidewalk and began hitting him in the head. By the time a passing truck driver separated the two men, McHan was unconscious, his nose broken.

McCullough's trip from Atlanta to the State Prison Farm in Milledgeville must have seemed ironic, for that was where his father had lived throughout McCullough's childhood. Father and son just missed each other there; Andrew had been paroled earlier in 1929. But he returned before Roy's sentence was up — this time on the way to the electric chair. Gov. Eugene Talmadge had denied the final appeal of his murder conviction in the death of William B. Baker, one of Georgia's best-known businessmen and one of Fayette County's best-loved citizens. Chairman of the Georgia Manufacturers Association, Baker had retired in 1930 as president of Atlantic Ice and Coal Company in Atlanta and taken up peach farming in Fayette County.

On September 25, 1932, Andrew McCullough and his son Alvin had spent a warm Sunday afternoon on the banks of the Flint River, fishing and drinking. As they headed home in the late afternoon, they passed near a cluster of small wooden houses, the homes of black tenant farmers who worked William Baker's 900 acres outside Woolsey.

"The McCulloughs had all the Negroes out there hollering and dancing, shooting at their feet," recalled Weldon Stubbs, a Woolsey resident who heard the pistol shots.

According to a newspaper report at the time, one of the black tenant farmers — Joe Redding — had testified against McCullough in the Spalding County trial stemming from the murder of Walter Reeves. That had been twenty-one years earlier, but McCullough, now fifty-seven, supposedly had sworn vengeance against Redding, a relative of the Mitch Redding McCullough had wounded in 1911.

Redding barricaded himself in his home and sent one of his children to notify Baker. As soon as Baker and his son, Dr. W. Pope Baker, arrived at the tenant houses and began to approach the McCulloughs, Andrew opened fire with his pistol. One bullet passed through Baker's abdomen; another hit him in the head. McCullough continued firing, forcing the two men to return to their car and drive away under a barrage of shots. The elder Baker died en route to a hospital in Atlanta.

Roy, who was finishing up his sentence for the assault of McHan, was taken to his father's cell the day before the execution. They spoke through the bars only briefly. It was their first meeting in twenty-one years.

The next morning — August 25, 1933 — Andrew told his attorney to ask his two daughters not to "think hard of the boys," referring to his six sons, three of whom had run afoul of the law. He then confessed, "Whiskey did it. I had been drinking considerably that day, and what actually happened is not at all clear to me."

Roy was released from the State Prison Farm later that year and went right back to bullying and stealing in Atlanta. He was given six months on the chain gang for assault and battery in 1934 and then eighteen months to three years in 1935 for stealing a Plymouth Coupe. By the time he was paroled in summer of 1937, at age twenty-five, he was a hardened criminal. He was back in state prison within two years. During the winter of 1939, working with three other men and a woman named Dollie Johnson, he burglarized a grocery store and took, among other things, two shotguns. He was caught two weeks later when he had a wreck while driving drunk and left the scene of the accident. He was also charged with having stolen a Ford V-8 Coach five days after the robbery. He pleaded guilty to both charges and was sent to the new state prison at Reidsville.

It was at Reidsville, Ray Maddox says, that he met McCullough.

"He [McCullough] could be your best friend and, in his good moments, had a likeable personality, but don't get in his way or in any way do him a hurt, for if you did he wouldn't rest until you were a dead man," writes Maddox in his autobiography. "While in prison he killed two or three guards and four or five fellow inmates at various times. Nobody could be better to you than Candy, if he liked you, and nobody was more dangerous than he was if he hated you. . . . I don't know that I'll ever understand how Candy could be a warm friend one minute and totally heartless the next, void of any basic human emotions."

Otto Spencer, who served time with McCullough and Maddox at Ben Hill, described McCullough during a 1986 interview: "He wasn't scared of the devil; he was always smiling. When he was in prison, I never remember him having a visitor. I think that's why he went so bad. I never heard him mention any family."

After McCullough's parole in 1942, the "Georgia Boys" began robbing gamblers and crooks. Maddox says in his autobiography their first job was at Harper's Cherokee Club, a gambling house in Macon. He says he, McCullough, and three other men rushed in with drawn guns, lined everyone against the wall, and took their money. They collected more than $5,000 and headed back to Atlanta, convinced nothing could be easier. Maddox says other out-of-town jobs followed, with McCullough always back in Atlanta in time to check in with his parole officer.

In November 1943, McCullough got twenty days in jail for being drunk. In September 1944, he was sentenced to eighteen months for stealing a car and was sent to Ben Hill. After the camp burned, McCullough finished out his sentence and was not arrested again until April 1946, when he got thirty days for drunk driving. He escaped twice within a week but was recaptured and sentenced to six months in state prison. He was back on the streets of Atlanta October 7, ready to embark on a series of robberies.

On Halloween, McCullough robbed the Best Liquor Store on Marietta Street. The next day, he and Joe Lee Bishop used "force and intimidation," according to police reports, to take fifty-six dollars and a gold watch from Louis Triff and sixty-seven dollars from E. I. Fox. Then on November 26 — ten days before the Winecoff fire — McCullough and Bishop held up T. M. Waters' grocery store on Murphy Avenue. They took a bag containing $1,408 from Waters, as well as the wallets of two other men in the store, than locked them all in a walk-in refrigerator. Three days later, McCullough, Bishop, and seventeen-year-old Bill Davis pulled off the Pig 'n' Whistle holdup. They had followed Nell Winters to the restaurant from a bank where she had drawn out the money to pay the employees their weekly wages, then forced her and C. D. Lemming to lie on the floor, picked up the two payroll bags, and walked out to a waiting car. Within an hour, Bishop and Davis had been picked up on suspicion of robbery.

When McCullough was arrested the day after the holdup on suspicion and again the day of the fire, he listed his address as 391 Currier Street, located in a segregated black neighborhood northeast of the Winecoff. But McCullough apparently had been spending time at 117 Richardson Street, one block from Formwalt Street south of downtown. Two days after the fire, detectives swept down on the house and busted Atlanta's "worst dope ring in years," confiscating a large quantity of morphine. Police said they began watching the house when told one of the Pig 'n' Whistle robbers had taken refuge there. Because Bishop and Davis had been arrested at Bishop's residence on Boulevard immediately after the holdup, McCullough had to be the suspect hiding out on Richardson Street.

Upon being convicted after the fire on four counts of robbery, McCullough was sent to Georgia's toughest prison, the Buford Rock Quarry Prison for Incorrigibles, later made infamous by an Atlanta *Journal* investigative report that inmates there broke their legs with sledgehammers in order to be transferred.

McCullough also served time at Reidsville, the Wayne Prison Branch near Jesup, and the Lee Prison Branch near Leesburg — the latter two being among the seven prison branches that provided manual labor across the state. Maddox says McCullough escaped several times during his sentence, but there are no records available for substantiation. It's possible that McCullough never spent a day out of prison — whether by design or coincidence — after December 7, 1946.

Georgia Department of Corrections officials said in 1992 that McCullough's prison file was destroyed according to standard practice for old inmate records. They say the only record of McCullough's imprisonment is an index card showing where he was incarcerated.

Maddox claims investigators went to prison to question McCullough about the fire. "They suspected him but they didn't have any proof," said Maddox in 1992.

Maddox's account of how the fire started has previously gone unpublicized for several reasons. The foremost is that his paperback autobiography was not widely distributed and is now out of print. And not only has McCullough been long dead, but so has J. H. "Pat" Morgan, who Maddox says was the teenage partner in crime accompanying him and McCullough the night of the fire. Morgan died April 30, 1958, in a much-publicized murder-suicide at a Gulf service station next to Ponce de Leon Park, where he had been arrested in 1954 as one of the "flyball" gamblers. Police said Morgan, who had a long police record, shot his estranged wife, Martha, and then turned the gun on himself.

The possibility also exists that Leo Nahlik used his wide influence to protect McCullough, if he did indeed rob the poker game in Room 330 at Nahlik's direction.

The son of a Bohemian immigrant who moved to Atlanta in 1923 as a tailor, Nahlik is remembered both as an outstanding detective and as a kingpin of vice. He was a strikingly handsome man who favored silk ties, tailor-made suits, and diamond rings. A onetime boxer, he had been a photographer for the Atlanta *Georgian* before becoming a policeman. He loved publicity and was always quotable, a police reporter's dream. He was seemingly fearless as well.

"Leo was probably the best detective the Atlanta police ever had," said longtime Atlanta *Constitution* police reporter Keeler McCartney in 1992. "It was a game to him to see if he could catch somebody. After he caught them, he sort of lost interest. He didn't really care about trying the case."

Melvin Coppenger, Nahlik's police partner from 1935 to 1942, said he and Nahlik made more arrests one year than the whole department. "Whatever he did after work, he did a fine job during work," he said. "But I never had any indication he was doing anything. He was a ladies' man, I guess. Hookers were the only class of people who would have anything to do with Leo that I know of. I mean the word was out all over town. Maybe Leo was a contact person for prostitutes. He called them broads."

"The rumor was," said McCartney, "that Leo slept with all the prostitutes and gambled with all the gamblers, so he knew everything that was going on."

"Leo was at home with people who were on the edge of criminality," said former Atlanta police detective Lamar Moseley in 1993. "When he left [the force], it immediately got out that he was into something else. Now he could've been involved in anything while he was working — he could've been — but I never heard of it."

Coppenger and Helms both remember that Nahlik was under investigation when he abruptly left the police force in 1948. Coppenger said Nahlik became a professional gambler afterward. Helms and others said he ran a gambling operation at a boathouse on Lake Jackson, south of Atlanta. Later he lived on a farm between Griffin and Jackson. It was there, on the night of February 4, 1957, that he murdered his wife.

Floy Estelle Nahlik, his second wife, was twenty-three years younger than Nahlik. He told the Spalding County sheriff that he found her with James Hollis, a sixteen-year-old black laborer on Nahlik's farm, and shot them both with a twelve-gauge shotgun. Hollis was killed; Floy Nahlik was critically wounded and died twelve days later. Nahlik told the sheriff he had been taking medication that made him sleep a lot, and on that evening he awoke to discover Hollis and his wife together in the living room. The shooting received front-page headlines in the Griffin *Daily News*, but charges were never filed against Nahlik, and the grand jury refused to indict him.

Nahlik apparently remained active in gambling and the lottery in the 1960s. Atlanta police chief Herbert Jenkins included Nahlik in a 1965 departmental bulletin listing six men suspected of being both bookies and lottery bankers. Nahlik died in Atlanta in 1969 and was buried at Marietta National Cemetery.

McCullough had died five years earlier while serving a life sentence in prison.

On August 17, 1959, he was working on a detail pulling up pine seedlings at Herty Nursery, a state-operated tree nursery that the Lee Prison Branch inmates helped maintain in nearby Albany. Also working at the nursery on that steamy south Georgia day was sixty-four-year-old Albany carpenter W. T. Winslette, who was completing the cabinet work on a home being constructed for the nursery superintendent. Winslette had recently been ill and therefore was without a crew, so he had several trusties working with him. He'd often bring them a blueberry pie.

As was his daily custom at lunchtime, Winslette changed clothes to go into town to eat. McCullough, who had been watching this routine, showed up at the house and demanded Winslette's keys. When Winslette refused to give them to him, McCullough picked up a two-by-four, staggered Winslette with a blow to the head, then stabbed him in the ribs with a homemade knife.

McCullough took Winslette's keys but couldn't start the car. He then flagged down a nursery worker in a truck and asked him to take him out of the nursery, but the truck driver refused, denying McCullough any hope of escape. McCullough found another inmate, told him what he had done, and said, "Go tell somebody."

The police reported that when they arrived McCullough did not seem "quite right." He later admitted he had been drinking a soft drink mixed with paint thinner.

McCullough was brought to trial within six weeks, and testimony lasted only one day. The jury deliberated less than two hours. "They didn't go for the death penalty. They said he'd never get out of prison anyway," Tucker McDonald, Winslette's son-in-law, said in a 1992 interview.

McCullough returned to Reidsville with a life sentence. He died there less than five years later, on January 6, 1964, after having been in the prison hospital for twenty-four days. The cause of death was listed as "esophageal varices," usually caused by alcoholism or high blood pressure.

Maddox says in his book that McCullough was poisoned. But both Perry Moore, who performed the autopsy, and Joseph Arrendale, surgeon and head physician at the prison hospital then, said in 1991 that they did not remember anything about McCullough or his death. According to Department of Corrections officials, McCullough's hospital records would have been destroyed with his prison file.

McCullough's body went unclaimed by his family and, after being advertised a week, was buried in Grave 202 of the prison cemetery — a field of wooden crosses marked only by a prison ID number, about a mile from the front gate.

Within a couple of years of McCullough's death, a young Atlanta newspaperman named Gene Moore began working on a book about the Winecoff fire. By day he would write editorials for the Atlanta *Journal*, and on nights and weekends he would interview the dozens of people whose lives had been touched tragically by the Winecoff or who had worked on the Winecoff case. The book was the idea of *Constitution* editor Ralph McGill, who thought Moore — not yet twenty-five — had great potential and that the Winecoff story was strong material.

Moore's considerable research convinced him the fire had been set. He was believed the arsonist was involved in either the poker game in Room 330, the fight in 512, the Chicago underworld, or a romantic tryst.

Then one night in 1966, Moore's notes — which he kept in his *Journal* editorial office on Forsyth Street — disappeared. He figured there were at least thirty pounds of them, so he knew they had not been simply misplaced. But neither company security officers nor the Atlanta police ever turned up any clues. He never had the heart to restart his research.

EPILOGUE

━━✦━━

In the building where 119 people had died less than five years earlier, more than 500 people turned out for the grand opening of
the Peachtree on Peachtree Hotel on April 15, 1951. They toured
a hotel that had fire escapes, sprinklers, and even fire doors.

Ironically, the new owner was Fred Beazley, the man hand-picked
two decades earlier to be president of Atlantic Ice and Coal, succeeding
the retiring W. T. Baker, the Atlanta businessman murdered by Roy
McCullough's father.

The son of a railroad engineer, Beazley was a self-made man, an
entrepreneur who started a coal and wood delivery business at age sixteen in his native Portsmouth, Virginia. He made a million dollars by
the time he was twenty-eight. He lost that much in the stock market
crash of 1929 but promised his creditors that he would someday return
their money. He headed for Atlanta to rebuild Atlantic Ice and Coal, a
publicly held company with ice plants, cold storage facilities, and brew-
eries around the Southeast. A decade later, he opened his own chain of
ice and cold storage plants in the Southwest.

Beazley was a simple man and a teetotaler who cherished his 100-acre farm in Chamblee. Tight-fisted and strong-willed, he thrived on taking over a failing business and making it a success. He saw that opportunity in a building where catastrophe had occurred — as well as the opportunity to get into the booming hotel business at a bargain price. He paid the Arlington Corporation $700,000 in cash and then poured $1 million into renovation. His intention was for the Peachtree on Peachtree to be "the finest and safest hotel in the Southeast."

Each floor was given three means of exit: shaft-enclosed elevators, a fire escape, and a cement stairway constructed in the southwest corner, where the "28" rooms had been. The stairwell's metal fire doors completely isolated each floor from flames above or below it.

From his office on the top floor, Beazley ran a financially successful business, again beating the odds by overcoming the Winecoff fire's infamous stigma. Although many people would not even consider staying in a building where so many had died, the Peachtree on Peachtree attracted a regular clientele of businessmen. One of them was Michael Conners, a salesman from Montgomery, Alabama, who had crawled across a ladder from 1418 to the Mortgage Guarantee Building on December 7, 1946.

By the late 1950s, however, modern motels began redrawing the landscape of the American lodging industry. They were popping up everywhere, including downtown Atlanta. In 1956, the Heart of Atlanta opened only four blocks from the Peachtree on Peachtree. In 1963, Beazley moved back to Portsmouth and became less inclined to invest money in the hotel. Business inevitably deteriorated. When Beazley put his hotel on the market, there were no takers.

In 1967 — the year John Portman's Hyatt Regency opened with its atrium architecture three blocks up Peachtree Street — Beazley gave his hotel to the Georgia Baptist Convention to establish a ministry for the aged on the site where so many had died. The building became a retirement home: the Peachtree on Peachtree Inn.

Then in 1971, on Christmas Day, the Winecoff lost the distinction of being the site of the worst hotel fire in the world. On that day 162 people were killed in the Hotel Taeyokale in Seoul, South Korea.

But the horror of the Winecoff fire did not go away quickly. After leaving Atlanta, the Geeles — the Winecoff's operators — ran the Stonewall Jackson Hotel in Staunton, Virginia, for a short time before returning to the Gilcher Hotel in Danville, Kentucky. At the time of the Winecoff fire, they still operated the Gilcher as well as the Central Hotel in Macon, but those holdings were never touched by the court in compensating the fire's injured and the families of its victims. Financially, the Geeles had fared as well as possible. Not so, emotionally.

Art Sr. died less than thirty months after the fire. Art Jr. built Danville's first motel on U.S. Highway 27, the main route from Detroit to Florida, but he rarely talked about the fire. He felt immensely responsible, according to family members, and battled depression for the rest of his life. He died still believing the fire was set by an arsonist.

Fred Grennor, the building engineer on his second night on the job, died an alcoholic. "I don't know if the fire contributed to the alcoholism or not," said his son, Merrill, "but I know it really affected him for several years." Before starting an inspection of the hotel, Grennor had stopped by Suite 510-12 with bellhop Bill Mobley and was trapped there by the fire.

In 1949 Hilda Morrison summoned enough courage to climb the stairs to 1026, the room where her husband, Dwight, had died. "You think you want to go up there," she said, "but then you get up there and it's like you don't have any legs."

In 1986, forty years after the fire, the pain was still overwhelming for seventy-year-old Bob Knox. He broke down twice while talking about his parents' death during an interview for a fortieth anniversary newspaper article. His father, Peter Knox, was a prominent, seventy-eight-year-old businessman and civic leader in the east Georgia town of Thomson. His wife, Gertrude, had to come to Atlanta to see an ear specialist. They suffocated in each other's arms on the bed in 1218.

Ruth Minnix — the Columbus mother who tried to keep her fifteen-year-old son, Rutledge, home and then, once in Atlanta, attempted to convince him to spend the night with her at the Henry Grady — told the reporter: "I think those were the most horrible hours of my life. I would have a thousand times rather been in that room than him." Then, at the end of the interview, she remarked, "It's nice to have someone remember him."

Forty-seven years after the fire, the Winecoff is rarely remembered. Many of the thousands who pass the building everyday on foot, in a car, or on a bus, are not aware that it was the site of what is still the worst hotel fire in U.S. history — the MGM Grand Hotel fire in Las Vegas in 1980 being a distant second with eighty-four victims.

But there it sits, its tall, skinny, square shape incongruously set down on Peachtree Street's most impressive four-block stretch, glittering with skyscrapers of glass and steel. The Davison-Paxon's building is still across Ellis Street, but it's now called Macy's. The Mortgage Guarantee Building is now the Carnegie Building, still ten feet away, across a dead-end alley that seems frozen in memoriam. Otherwise, the new neighborhood hardly resembles the old. The Henry Grady is gone; so are the Capitol and the Roxy. The Georgia Pacific tower looms over the site of the Paramount and Loew's Grand, the latter destroyed by fire on a freezing January day in 1978. An investigation revealed that it was arson, but no arrests were ever made.

The big Coca-Cola clock, Lane's Drug Store, and the Frances Virginia Tea Room are as gone from Peachtree Street as the trolley tracks that lined it in 1946. An underground MARTA rail station now occupies the half-block south of the Winecoff where once stood Chandler's Shoe Store, Jack Sheriff's Theatre Restaurant, and Hamburger Heaven. Passengers emerging from the station can look up and see the fire escape installed on the hotel's south side before it reopened in 1951. It's outside the windows of the doomed "30" rooms, exactly where it was needed the most in 1946.

Thousands sleep nightly in the 4,500 hotel rooms in the Winecoff's old neighborhood, proof that the Winecoff's legacy has hardly impeded high-rise hotel building in Atlanta, a city that now lives by its tourism and hospitality industry. The twenty-five-story Ritz Carlton is diagonally across the street. Where the Henry Grady once stood, the seventy-three-story, circular Westin Peachtree Plaza — the tallest hotel in the world — now shadows the Winecoff like some giant bully. The Hyatt Regency is just up the street, and the forty-seven-story Marriott Marquis is only four blocks away.

Meanwhile, the struggling retail sites on the ground floor and mezzanine of the building have offered the only semblance of life since the Georgia Baptist Convention sold out to Ackerman & Company for $2 million in 1981. The company, which had purchased the old Mortgage Guarantee Building in 1979, wanted to combine the two buildings for office use. But in 1983 Ackerman announced that the Winecoff building would be torn down and a multi-story office complex would arise in its place. Those plans fell through also, and by 1985 an agreement had been reached with a Minneapolis developer who intended to turn the Winecoff into a sixty-five-room luxury hotel with suites on the upper floors and retail shops on the lower ones. The deal collapsed in the financing stage.

Just as the size of the building enabled the Winecoff to be built without fire escapes, the smallness of the lot has deterred potential developers, as has the fact that the site is simply saturated in tragedy.

Finally, in 1990, Ohio Savings Bank, holder of the $2.1 million mortgage on the property, foreclosed after Ackerman tried to renegotiate terms that weren't acceptable. The building was offered at auction on the Fulton County Courthouse steps on — appropriately — December 7, 1990. There were no takers. Ohio Savings Bank retained ownership and announced in late 1992 that it had four offers at an asking price of $2.2 million, a bargain on Peachtree Street. Possible uses listed by prospective buyers included condominiums, apartments, a hotel, or a dormitory for Georgia State University.

But in 1993 the Winecoff still remains a sort of haunted house on Peachtree, beige paint peeling from its windows on the outside and plaster littering the floor of the small, empty rooms inside. The China Star Express restaurant sits on the ground floor, and there's a hair salon on the former mezzanine, but nowhere are there any reminders that a fire ever took place at the corner of Peachtree and Ellis — not even a memorial marker for the 119 people who died there.

ACKNOWLEDGMENTS

T he authors would like to express their appreciation to all those who gave of their time and cooperation in helping to make this book possible — the families of the victims, the survivors, and the firefighters.

In addition, the authors would like specifically to thank L. P. "Pat" Patterson, Bill Dodd, Tom Jones, Reid and Cary Horne, Hamilton Lokey, George Goodwin, Nell McDuffie, Arnold Hardy, Franklin Garrett, Herbert Jenkins, Keeler McCartney, William Hamer, Ann Howell, and John Hall.

Unless otherwise noted, photographs are courtesy of the Atlanta *Journal-Constitution*.

APPENDIX

--- ≡✦≡ ---

Victims of the Winecoff Fire:

Earlyne Adams, 16, Thomaston, Ga., Room 1430. Youth Assembly delegate. Senior at R.E. Lee Institute.

Sarah Aldridge, 28, Eva, Ala., Room 1028. Working internship at Rich's. Junior at University of Alabama. Worked with USO during World War II. Leader in women's intramurals program.

Florence Baggett, 43, Claxton, Ga., Suite 1108-1110-1112. One of best known auctioneers in Southeast. Came to Atlanta to see doctor about enlarged vocal chords.

Walter Baker, 41, Asheville, N.C., Room 1102. Vice president of Asheville Truck and Equipment Co. In Atlanta to buy war-surplus vehicles. Father of two children.

Clarence Bates, 14, Bainbridge, Ga., Room 924. Youth Assembly observer. Sophomore at Bainbridge High School. Member of school band, basketball team. Boy Scout.

Walton Beck, 47, Jacksonville, Fla., Room 1226. Owner of Beck Oil and Equipment Co. In Atlanta on business. Spent week at Winecoff.

Bill Berry, 25, Cedartown, Ga., Room 1626. Former Marine. Survived Guadalcanal. Bartender at Jack Sheriff's Theater Restaurant, next to Winecoff. Semi-professional bowler.

Sue Broome, 16, Bainbridge, Ga., Room 920. Youth Assembly reporter. "Most Popular" in Bainbridge High School senior class. President of glee club. Frequent church soloist. Cheerleader.

Lamar Brown, 16, Rome, Ga., Room 1030. Youth Assembly delegate. Senior at Rome High School. City badminton champion. Avid stamp collector.

Louise Brown, 46, Cornelia, Ga., Room 1628. In Atlanta for doctor's appointment.

Bill Bryson, 28, Asheville, N.C., Room 928. Bus driver for Smoky Mountain Trailways. Staying in room rented permanently by bus company. Sailor during World War II. Father of two young daughters.

Eloise Buck, 40, St. Louis, Mo., Room 1124.

Edith Burch, 20, Chattanooga, Tenn., Room 1008. Married eight months. On "second honeymoon" weekend. Native of Cumberland, Ky. Husband, George, survived fire.

A.J. Burns, 26, Ossining, N.Y., Room 1416. Grandson of founder of Burns detective agency.

Peggy Cobb, 21, Fayette, Ala., Room 828. On internship at Davison-Paxon's. Senior at University of Alabama. Member of newspaper staff and Kappa Delta sorority.

Bill Cochran, 47, Miami, Fla., Room 1006. Plymouth salesman in Coral Gables. Visiting friends in Atlanta. Wife, Jake, survived fire.

Freida Constangy, 24, Atlanta, Ga., Room 1222. Staying at Winecoff until utilities turned on and furniture delivered at recently rented apartment. Three months pregnant.

Joe Constangy, 28, Atlanta, Ga., Room 1222. Husband of Freida Constangy. Army veteran. Had recently bought Sims Five and Ten Cent Store.

Morris Constangy, 55, Atlanta, Ga., Room 1528. Father of Joe Constangy. Son of Russian immigrants. Co-owner of dress factory. Moved into Winecoff permanently in September. Wife, Annie May, survived fire.

Elmer "Spike" Conzett, 32, Dubuque, Iowa, Room 1230. Navy lieutenant commander and war hero. Sank Japanese destroyer at Guadalcanal as dive-bomber pilot. Making one-night stopover in Atlanta while ferrying Navy planes from Norfolk, Va., to Long Beach, Calif.

Billie Cox, 31, Murphy, N.C., Suite 1002-1004. Came to Atlanta to go Christmas shopping and take 3-year-old son to see Santa Claus. Son survived fire and, according to mother's final wish, was raised by father's sister in Kansas and grew up to be a physician.

Bob Cox, 33, Murphy, N.C., Suite 1002-1004. Husband of Billie Cox. Physician. Former staff doctor with TVA at Hiwassee and Fontana dams construction sites. Chairman of hometown's upcoming polio fund drive.

Philip David, 26, Savannah, Ga., Room 1422. Air Corps gunner in World War II. Son of French immigrants. Native of Long Island, N.Y. Married Georgian. Came to Atlanta to prepare to enroll at Georgia Tech in 1947.

Mary Alice Davis, 48, Bainbridge, Ga. Room 926. Advisor to Youth Assembly delegation from Bainbridge High School. Youth director at First Methodist Church. Graduate of Huntingdon College.

Matsey Dekle, 69, Tampa, Fla., Room 1402. President of Dekle Holding Co. Native of Thomasville, Ga. Staying in Atlanta while daughter and son-in-law visited new summer home in Gatlinburg, Tenn.

Mary Dickerson, 34, Douglas, Ga., Room 1630. Supervised government-run nursery during World War II. Graduate of Florida State College for Women. Spending weekend with husband while awaiting completion of new house.

Will Dickerson, 38, Douglas, Ga., Room 1630. Husband of Mary Dickerson. Director of unemployment compensation for Georgia's labor department. Graduate of Washington and Lee law school. Survived Battle of Bulge.

Mary Melinda Dickerson, 6, Douglas, Ga., Room 1630. Daughter of Will and Mary Dickerson. Had started first grade in September.

Bill Dickerson, 3, Douglas, Ga., Room 1630. Son of Will and Mary Dickerson.

Eric Ellicott, 46, Jacksonville, Fla., Room 1414. National sales manager for service merchandising company headquartered in Atlanta.

Robert Fluker, 41, Thomson, Ga., Room 702. Safety engineer for J.A. Jones Construction Co. Georgia Tech graduate. Stopped in Atlanta for dentist appointment before going overseas to rebuild Okinawa air base.

Gladys Goodson, 29, Clay City, Ill., Suite 1202-1204. Native of Grand Rapids, Mich. Met husband when he was Gulf oil scout there. Had reservations at another hotel but room was given to someone else.

Joe Goodson, 35, Clay City, Ill., Suite 1202-1204. Husband of Gladys Goodson. Geologist for Pure Oil Co. Graduate of University of Kentucky. En route to Florida to recover from pneumonia.

Joe Goodson Jr., 4, Clay City, Ill., Suite 1202-1204. Son of Joe and Gladys Goodson.

Barbara Goodson, 1, Clay City, Ill., Suite 1202-1204. Daughter of Joe and Gladys Goodson.

Earl Carl Gragg, 14, Bainbridge, Ga., Room 924. Youth Assembly observer. Sophomore at Bainbridge High School. President of Junior Hi-Y. Pianist for church youth group. Boy Scout.

Patsy Griffin, 14, Bainbridge, Ga., Room 926. Youth Assembly observer. Sophomore at Bainbridge High School. Daughter of Georgia adjutant general — and future governor — Marvin Griffin.

Frank Hale, 31, Nashville, Tenn., Room 1426. Machinist for United Shoe Machine Corp. In Atlanta on business. Father of 4-year-old daughter.

Julia Hall, 16, Albany, Ga., Room 822. Youth Assembly delegate. Senior at Albany High School. Student council member. Homeroom president.

Lena Harris, 35, Ducktown, Tenn., Room 824. Bank cashier. In Atlanta to go to theater, shop, and visit sister, Irene Tollett.

Christy Hinson, 17, Thomaston, Ga., Room 1430. Youth Assembly observer. Senior at R.E. Lee Institute. Scheduled to enter Agnes Scott College in 1947.

Harold Irvin, 23, Cornelia, Ga., Room 830. Torpedo bomber pilot in Pacific during World War II. Stopped in Atlanta to visit friends and celebrate recent Navy discharge. About to start job as Eastern Airlines pilot.

John Irwin, 23, Long Island, N.Y., Room 422. Salesman for the Luxene Corp. Father of 4-month-old baby.

Jerry Jenkins, 15, Donalsonville, Ga., Room 930. Youth Assembly delegate. Senior at Seminole High School.

William F. Jones, 21, Oak Ridge, Tenn., Room 818. Clerk typist for U.S. government's Manhattan Project in Oak Ridge. Came to Atlanta for weekend with co-worker.

Irene Justice, 27, East Point, Ga., Room 624. Worked for firm in nearby Candler Building and attended law school at night

Charles Keith, 17, Rome, Ga., Room 1030. Youth Assembly reporter for Rome High School. Senior class historian. Older brother first person from Rome to die in World War II.

Gertrude Knox, 68, Thomson, Ga., Room 1218. In Atlanta for doctor's appointment.

Peter Knox, 78, Thomson, Ga., Room 1218. Husband of Gertrude Knox. One of Thomson's leading citizens. Longtime member of both city council and school board. Retired from successful lumber business.

Grace Lain, 52, Birmingham, Ala., Room 1428. Former director of youth extension work for Alabama Baptist Convention.

Paul Lain, 54, Birmingham, Ala., Room 1428. Husband of Grace Lain. In Atlanta on business. Sold surplus goods through the War Assets Administration.

James Little, 25, Charlotte, N.C., Room 1514. Graduated from Georgia Tech in 1944 and went to war. Officer on Navy destroyer. Won 11 battle stars and Presidential citation. In Atlanta for sales training with Firestone.

Eloise Mason, 59, Atlanta, Ga., Room 1104. Retired stenographer. Staying at Winecoff while looking for permanent residence.

Jacob Mauss, 41, Philadelphia, Pa., Room 1424. In second week at Winecoff. Father of two children.

Borgia McCoy, 58, Portland, Maine, Room 724. Mother-in-law of British vice-consul Thomas Bolton. Living in Atlanta to help care for grandchild while daughter worked. Staying at Winecoff until new home ready.

Gwen McCoy, 17, Gainesville, Ga., Room 1130. Youth Assembly delegate. Senior at Gainesville High School.

Jimmy McDonald, 39, Memphis, Tenn., Room 1624. CPA for Federal Public Housing Authority. Stationed in Atlanta but traveled throughout Southeast. Wrote home to wife and daughter in Memphis every day.

Hedy Metcalf, 15, Columbus, Ga., Room 730. Youth Assembly delegate. Senior at Baker Village High School.

Sara Baggett Miller, 34, Gordon, Ga., Suite 1108-1110-1112. Widow and mother of two children. Made bags for kaolin mining company. Came to Atlanta to shop with sister, Catherine McLaughlin, who survived fire.

Rutledge Minnix, 15, Columbus, Ga., Room 1116. Youth Assembly delegate. Junior at Columbus High School. Member of student council and debate club.

Laura Miranda, 36, Atlanta, Ga., Room 826. Beautician in Loew's Grand Building, across from Winecoff. Had recently become grandmother.

Ella Sue Mitchum, 16, Gainesville, Ga., Room 1130. Youth Assembly delegate. Senior at Gainesville High School.

Mary Minor, 31, Thomaston, Ga., Room 1430. Advisor to Youth Assembly delegation from R.E. Lee Institute. Organized Tri-Hi-Y chapter there in 1945 in response to student requests. Mother of 4-year-old daughter.

Ralph Moody, 32, Arlington, Ga., Suite 1210-1212. Operator of garage. Came to Atlanta to see about buying sawmill equipment. Accompanied by friend, Billy Newberry, who survived fire.

Suzanne Moore, 16, Gainesville, Ga., Room 1130. Youth Assembly delegate. Senior at Gainesville High School.

Dwight Morrison, 26, Lumpkin, Ga., Room 1026. Pilot of B-26 nicknamed "Tobacco Road" during World War II. Flew 65 bombing missions over Europe. In Atlanta looking for job. Father of son born two months after fire.

Robbie June Moye, 16, Donalsonville, Ga., Room 930. Youth Assembly delegate. Treasurer of Seminole High School senior class. Cheerleader.

Mary Lou Murphy, 14, Bainbridge, Ga., Room 926. Youth Assembly observer. President of Bainbridge High School sophomore class.

Margaret Nichols, 30, Atlanta, Ga., Room 720. Press agent and hostess at Jack Sheriff's Theater Restaurant, next to Winecoff. Former Miss Atlanta runner-up. Nicknamed "Luscious."

Margaret Parker, 12, Montgomery, Ala., Room 1604. In Atlanta with mother to shop. Mother survived fire.

Tona Perry, 65, Tampa, Fla., Room 1402. Accompanied Matsey Dekle to Atlanta. Native of Tennessee.

Ed Pettyjohn, 41, Birmingham, Ala., Room 1122. Salesman for Winchester

Repeating Arms Co. Father of two children.

Ruth Powell, 16, Bainbridge, Ga., Room 920. Youth Assembly delegate. Selected "Best Mannered" in Bainbridge High School senior class.

Jean Pruett, 21, Fort Payne, Ala., Room 1028. Working internship at Rich's. Senior at University of Alabama. Fifth grade teacher in 1945-46.

Freddie Louise Pruitt, 21, Atlanta, Ga., (room not known). Moved to Atlanta to find secretarial work after graduating from high school in Lenoir City, Tenn. Worked in downtown insurance office.

Carl Rasmussen, 49, Des Moines, Iowa, Room 1620. Small-town Midwest physician who joined Veterans Administration in May. Transferred to Atlanta in November. Staying at Winecoff until house ready. Scheduled to move out following week. Wife, Helen, survived fire.

Richard Rasmussen, 20, Des Moines, Iowa. Room 1620. Son of Carl and Helen Rasmussen.

Emelda Reeves, 21, Atlanta, Ga., (room not known). Secretary for wholesale liquor distributor. Separated from husband. Mother of 4-year-old son. Friend of Freddie Louise Pruitt, who also died in fire.

Gloria Riemann, 20, Gulfport, Miss., Room 728. Working internship at Davison-Paxon's. Senior at University of Alabama. Member of Delta Gamma sorority.

Jeannette Riley, 16, Donalsonville, Ga., Room 930. Youth Assembly delegate. Senior at Seminole High School.

Wylie Rochelle, 46, Marietta, Ga., 46. Traveling salesman for Sonnebarn Paint Co. Loved to go to Georgia Tech football games and Atlanta Crackers baseball games.

Marianne Scherle, 16, Albany, Ga., Room 822. Youth Assembly delegate. Senior at Albany High School. Assistant editor of yearbook. President of youth fellowship at church.

Jack Sheriff, 32, Atlanta, Ga., Room 720. Owner of Jack Sheriff's Theater Restaurant, a night club next door to Winecoff. Father of four children.

Nell Sims, 33, Barnesville, Ga., Room 1504. Secretary at Aldora Mills. President of Business and Professional Women's Club. Came to Atlanta at request of friend who was buying trousseau and meeting fiancee.

Cleve Sisk, 48, Asheville, N.C., Room 928. Bus driver for Smoky Mountain Trailways for 26 years. Staying in room rented permanently by bus company.

James "Buzz" Slatton, 16, Rome, Ga., Room 1030. Clerk of house for Youth Assembly. Senior at Rome High School. Member of band. Planned to be governor of Georgia.

Anne Smith, 14, Valdosta, Ga., Room 1502. On shopping trip with mother. Overcame 80 percent deafness sufficiently to attend public school.

Dorothy Kiker Smith, 36, Fitzgerald, Ga., Room 1530. Former court reporter for Cordele circuit. Came to Atlanta on shopping trip with sister, Boisclair Williams.

Fred Smith, 14, Fitzgerald, Ga., Room 1530. Son of Dorothy Kiker Smith.

Dotsy Smith, 12, Fitzgerald, Ga., Room 1530. Daughter of Dorothy Kiker Smith.

Mary Smith, 4, Fitzgerald, Ga., Room 1530. Daughter of Dorothy Kiker Smith.

Bobby Sollenberger, 16, Barnesville, Ga., Room 1126. Youth Assembly delegate. Junior at Milner High School. Member of basketball team and debate club.

Harry Sorrells, 47, Asheville, N.C., Room 928. Bus driver for Smoky Mountain Trailways. Staying in room rented permanently by bus company. Loved to tinker with machinery. Father of 2-year-old daughter.

Ethel Stewart, 20, Ralph, Ala., Room 1228. Working internship at Rich's. Senior at University of Alabama. Worked at father's country store. Engaged to be married in 1947.

Mary Stinespring, 58, Cornelia, Ga., Room 1628. Accompanied neighbor and close friend, Louise Brown, to Atlanta on shopping trip. Widow and mother of five children. Active church worker.

Harriett Strickland, 16, Iron City, Ga., Room 930. Youth Assembly delegate. President of Seminole High School senior class. Attended first Youth Assembly in March.

Elaine Sullivan, 20, Tuscaloosa, Ala., Room 1028. Working internship at Davison-Paxon's. Senior at University of Alabama.

Herbert Swanson, 38, Chicago, Ill., Room 522.

Ed Thomas, 55, Asheville, N.C., Room 1102. Owner of Thomas Buick and Asheville Truck and Equipment Co. In Atlanta to buy war-surplus vehicles. Scheduled to become Kiwanis Club president in January. Boy Scout leader.

Frances Thompson, 17, Gainesville, Ga., Room 1130. Youth Assembly delegate. Senior at Gainesville High School.

Bess Thruen, 51, Columbus, Ga., Room 1526. Wife of Charles Thruen. Recently bought house.

Charles Thruen, 51, Columbus, Ga., Room 1526. Husband of Bess Thruen. Civilian clerk in the salvage office at Ft. Benning. Recently retired from Army after 28 years of service.

Bill Todd, 39, Columbia, S.C., Room 1214. South Carolina director of Civil Aeronautics Administration. In Atlanta to attend meeting at regional headquarters. Father of daughter born six weeks after fire.

Irene Tollett, 44, Atlanta, Ga., Room 824. Librarian at Russell High School in East Point. Earned Ph.D. in library science in 1942. Spending night with sister, Lena Harris, who was visiting from Tennessee.

Virginia Torbert, 16, Thomaston, Ga., Room 1430. Youth Assembly reporter. Senior at R.E. Lee Institute.

Dot Tyner, 16, Columbus, Ga., Room 730. Youth Assembly observer. Senior at Baker Village High School.

Patsy Uphold, 16, Thomaston, Ga., Room 1430. Youth Assembly observer. Senior at R.E. Lee Institute.

Billy Walden, 16, Rome, Ga., Room 1030. Youth Assembly delegate. Senior at Rome High School. One of best football players in history of school. President of Hi-Y and Sunday school class.

Ann Walker, 16, Albany, Ga., Room 822. Youth Assembly delegate. Senior at Albany High School. President of Bookworm Club.

Helen Walker, 31, Atlanta, Ga., Room 528. Winecoff auditor. Vice-president of Women's Chamber of Commerce. Scheduled to graduate from Atlanta Law School in five days and go to work for judge in Metter, Ga.

E.B. Weatherly, 63, Cochran, Ga., Room 1024. Macon attorney who moved to farm during Depression and became one of South's top cattlemen. Former chairman of federally appointed committee on beef industry. Father of six children.

Boisclair Williams, 40, Cordele, Ga., Room 1520. On shopping trip with sister, Dorothy Kiker Smith.

Clair Williams, 8, Cordele, Ga., Room 1520. Daughter of Boisclair Williams. Wanted to come to Atlanta to see *Song of the South*.

Robert C. Williams, 20, Long Beach, Calif., Room 1230. Navy ensign. Was making one-night stopover in Atlanta while ferrying planes from Norfolk, Va., to Long Beach.

Maxine Willis, 16, Bainbridge, Ga., Room 920. Youth Assembly delegate. Senior at Bainbridge High School. Recently selected "Miss BHS" candidate.

Irene Wilson, 55, Knoxville, Tenn., Room 1224. Bookkeeper for Southern Coal and Coke. Lived with and cared for widowed mother. Came to Atlanta with two co-workers for weekend shopping trip.

Grace Winecoff, 76, Atlanta, Ga., Suite 1011-1012. Wife of hotel builder William Fleming Winecoff. Resident of hotel for 31 years.

William Fleming Winecoff, 76, Atlanta, Ga., Suite 1011-1012. Built Winecoff Hotel in 1913. Turned operation over to Robert Meyer chain in 1915. Sold hotel during Depression but continued to live there.

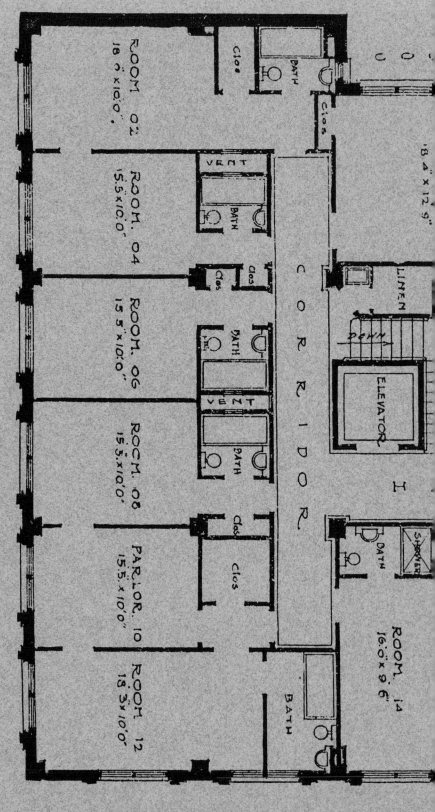